GIS IN WATER RESOURCES ENGINEERING

By
Gajraj Singh

2012

SBS Publishers & Distributors Pvt. Ltd.
New Delhi

ISBN 13 : 9789380090511

First Published in 2012

Published by:

SBS PUBLISHERS & DISTRIBUTORS PVT. LTD.
2/9, Ground Floor, Ansari Road, Darya Ganj,
New Delhi - 110002,
INDIA
Tel: 0091.11.23289119 / 41563911 / 32945311
Email: mail@sbspublishers.com
www.sbspublishers.com

Printed in India by Chaman Enterprises, New Delhi.

Preface

A Geographic Information System (GIS) can be defined as a system for entering, storing, manipulating, analyzing, and displaying geographic or spatial data. These data are represented by points, lines, and polygons along with their associated attributes. For example, the points may represent hazardous waste site locations and their associated attributes may be the specific chemical dumped at the site, the owner, and the date the site was last used. Lines may represent roads, streams, pipelines, or other linear features while polygons may represent vegetation types or land use. It describes the development of decision support system for evaluating the impact of floods resulting from dam and levee break/breaching based on a two-dimensional shock capturing unsteady conservative finite-volume model that uses a Digital Elevation Model directly as a computational grid. GIS-based decision support system directly reads simulation results and allows the user to interface these results with spatial socio-economic data to determine the probability of loss-of-life and urban and agricultural flood damage.

Topographic maps were digitized and a Digital Elevation Model (DEM) was generated to simulate lakes bottom topography and surroundings. The DEM was of a 20 m resolution and was interpolated in ArcInfo's Topograd's module using digitized contours, drainage networks and spot heights layers. The DEM-simulated lakes were then directly overlaid on their counterparts from the co-incident 2002 and 2006 sub images. Finally, a GIS-based scenario was established to predict the future surface areas, configurations and storage of these lakes. The sequence is based on remote sensing and regional stream flow data and was automated within a GIS-based computational program: Hydro spot. A GIS-linked numerical model described in this paper shows that the flood levels in the delta depend on the combined impacts of high river flows in the Mekong River, storm surges, sea level rise, and the likely, future situation of the Mekong Estuary resulting from the construction of dams as well as many other dams proposed throughout the remaining river catchment.

Author

Content

1

Water Resources

OVERVIEW

HYDROLOGIC IMPLICATIONS OF CLIMATE

Climate change is likely to alter the hydrologic cycle in ways that may cause substantial impacts on water resource availability and changes in water quality. For example, the amount, intensity and temporal distribution of precipitation are likely to change. Less dramatic but equally important changes in run-off could arise from the fact that the amount of water evaporated from the landscape and transpired by plants will change with changes in soil moisture availability and plant responses to elevated CO_2 concentrations. This will affect stream flows and groundwater elevations. This overview briefly summarizes potential impacts on the most important water resource elements.

PRECIPITATION CHANGES

Along with the projected future warming there will be changes in atmospheric and oceanic circulation, and in the hydrologic cycle, leading to altered patterns of precipitation and run-off. The most likely will be an increase in global average precipitation and evaporation as a direct consequence of warmer temperatures. Evaporation will increase with warming because a warmer atmosphere can hold more moisture and higher temperatures increase the evaporation rate. On average, current climate models suggest an increase of about 1%–2% per degree Celsius from warming forced by CO_2. An increase in global average precipitation does not mean that it will get wetter everywhere and in all seasons. In fact, all climate model simulations show complex patterns of precipitation change, with some regions receiving less and others receiving more precipitation than they do now; changes in circulation patterns will be critically important in determining changes in local and regional precipitation patterns.

CHANGES IN PRECIPITATION FREQUENCY AND INTENSITY

Many have argued that, in addition to changes in global average precipitation, there could be more pronounced changes in the characteristics of regional and local precipitation due to global warming. For example, Trenberth *et al.* hypothesized that, on average, precipitation will tend to be less frequent, but more intense when it does occur, implying greater incidence of extreme floods and droughts, with resulting consequences for water storage. Thus, the prospect may be for fewer but more intense rainfall–or snowfall–events.

CHANGES IN AVERAGE ANNUAL RUN-OFF

Run-off changes will depend on changes in temperature and precipitation, among other variables. Arnell used several climate models to simulate future climate under differing emissions scenarios. The study linked these climate simulations to a large-scale hydrological model to examine changes in annual average surface run-off by 2050.

They found that all simulations yield a global average increase in precipitation, but likewise exhibit substantial areas where there are large decreases in run-off. Thus, the global message of increased precipitation clearly does not readily translate into regional increases in surface and groundwater availability.

HYDROLOGICAL IMPACTS ON COASTAL ZONES

The IPCC Working Group II TAR identifies several key impacts of sea level rise on water providers located in coastal areas, including:

- Lowland inundation and wetland displacement,
- Altered tidal range in rivers and bays,
- Changes in sedimentation patterns,
- More severe storm surge flooding,
- Increased saltwater intrusion into estuaries and freshwater aquifers, and
- Increased wind and rainfall damage in regions prone to tropical cyclones.

WATER QUALITY CHANGES

Where stream flows and lake levels decline, water quality deterioration is likely as nutrients and contaminants become more concentrated in reduced volumes. Warmer water temperatures may have further direct impacts on water quality, such as reducing dissolved oxygen concentrations. Prolonged droughts also tend to allow accumulation of contaminants on land surfaces, which then pose greater risks when precipitation returns.

At the other extreme, heavy precipitation events may result in increased leaching and sediment transport, causing greater sediment and non-point source pollutant loadings to watercourses. Floods, in particular, increase the risk of water source contamination from sewage overflows, agricultural land, and urban run-off.

WATER STORAGE AND MANAGEMENT

An intensified hydrological cycle could make reservoir management more challenging, because there is often a trade-off between storing water for dry period use and evacuating reservoirs before the onset of the flood season to protect downstream communities. Reservoirs have been usually sized to handle a certain amount of stream flow variability, determined from a relatively short historical record. If the variability increases, reservoirs may be undersized to meet demands or adequately serve as flood protection. Thus, it may become more difficult to meet delivery requirements during prolonged periods between reservoir refilling without also increasing the risk of flooding. Earlier spring run-off from snowmelt is a likely manifestation of global warming. To the extent that adequate reservoir space is available, changing the operation procedures of reservoirs could mitigate some of these effects.

GROUNDWATER CHANGES

In many communities, groundwater is the main source of water for both irrigation and municipal and industrial demands. Generally there are two types of groundwater resources–renewable and non-renewable. Renewable groundwater is directly tied to near-surface hydrologic processes; it is thus intricately tied to the overall hydrologic cycle and could be directly affected by climatic change.

In many places, the overdraft of renewable groundwater aquifers occurs because the rate of withdrawal exceeds the rate of recharge. In fact, renewable groundwater supplies are often thought of being the same resource as surface water because they are so intertwined.

Thus, climate changes could directly affect these recharge rates and the sustainability of renewable groundwater. Non-renewable groundwater supplies are usually derived from deep earth sediments deposited long ago and so have little climatic linkage.

WATER DEMAND CHANGES

Future climate change could affect municipal and industrial water demand. Municipal water demand–especially for garden, lawn, and recreational field watering–is affected by climate to a certain extent, but rates of water use are highly dependent on water resource regulations and local user education. Industrial use for processing purposes is relatively insensitive to climate change; it is conditioned by technologies and modes of use. Demand for cooling water would be affected by a warmer climate because increased water temperatures will reduce the efficiency of cooling, perhaps necessitating increased source water abstraction to meet cooling requirements (or, alternatively, changes in cooling technologies to make them more efficient).

REGIONAL CHANGES

Although the preceding sections postulate some expected hydrologic changes from global warming, these generalizations will not be applicable in all places and at all times. Watson *et al.* examined the regional impacts of climate change, with a particular focus on assessing vulnerability. The report noted that more than 1 billion people do not have access to adequate water supplies, and that some 19 countries, primarily in the Middle East and Northern and Southern Africa, face severe water shortfalls. This number could double by 2025, in large part because of the increased demand caused by economic and population growth. Climate change could exacerbate the situation.

Watson *et al.* noted that many developing countries are particularly vulnerable to climate change because they already experience water shortfalls, being in arid and semi-arid regions. Many people derive their water from single-point systems, such as boreholes or isolated reservoirs.

This lack of "water diversification" increases people's vulnerability to water shortage. These systems do not have the redundancy necessary to minimize the risks during times of shortage. Also, given the often limited technical, financial and management resources available to developing countries, adjusting to shortages or implementing adaptation measures can impose a burden on national economies.

These small water supply systems are found in many parts of the world. Persistent drops in water level in these systems could adversely affect the quality of water by increasing the concentrations of sewage waste and industrial effluents, thereby increasing the potential for outbreaks of disease and reducing the quantity of potable fresh water available for domestic use.

There is evidence that flooding is likely to become a larger problem in many temperate and humid regions, requiring adaptation not only to droughts and chronic water shortages but also to floods and associated damage, raising concerns about dam safety.

Trenberth *et al.* hypothesized that global warming due to enhanced GHGs could increase the intensity of precipitation and reduce its frequency, which would be particularly problematic in regions with rapid changes in land use and land cover, because this would mean changes in surface run-off and groundwater recharge characteristics.

Flooding could be worse, accompanied by rapid drying and less overall water resource availability. The effects on water resources could be sufficient to lead to conflicts among users, regions and countries.

CLIMATE CHANGE INFORMATION FOR WATER RESOURCE MANAGEMENT

Water managers are accustomed to adapting to changing circumstances, many of which may be analogues of future climate change, and they have developed a wide range of adaptive options that may or may not be appropriate, depending on the magnitude of climate change and how rapidly and when those changes are put into affect.

For reasons noted above, climate warming will inevitably challenge existing water management practices, especially in countries with less experience in incorporating uncertainty into water planning and less financial and institutional resources. The current challenge is to incorporate climate change uncertainty along with the other types of uncertainty usually treated in the water planning process.

A cornerstone of climate change analysis in the water planning process is the use of hydrologic simulation to study the effect of a changing climate on the rainfall–run-off process. Many of these models attempt to capture the physical mechanisms of run-off production across the landscape by characterizing precipitation on to the land surface, either directly or through snowmelt, and the partitioning of that water into evapotranspiration, run–off to the river network, and recharge to groundwater systems.

Integrated Water Resource Management (IWRM) models can then use these water fluxes to determine reservoir management and water delivery strategies, often within a well-defined regulatory framework.

A review of the scientific and water planning literature, however, suggests that most water resource studies on climate change have incorporated climate information into their planning process using a top-down approach. This approach typically begins by establishing the scientific credibility of human-caused climate warming, develops future climate scenarios to be used at the regional level, and then imposes those potential changes on water resource systems to assess, for example, system reliability.

The problems with a top-down approach are as follows:

- It does not always address the unique needs of a region, and
- The approach may become mired in the uncertainty of the future climate projections.

Alternatively, the bottom-up approach begins by identifying a water sector's most critical vulnerabilities; articulates the causes for those vulnerabilities; suggests how climate change, climate variability, and climate extremes might or might not exacerbate those vulnerabilities; and finally designs an analytic process to better address and solve the vulnerability in the face of the climatic uncertainty (*e.g.*, a precautionary approach).

In either top-down or bottom-up approaches, IWRM can be the most effective method for assessing adaptation options and their implications in the context of an evolving regulatory environment with its competing demands.

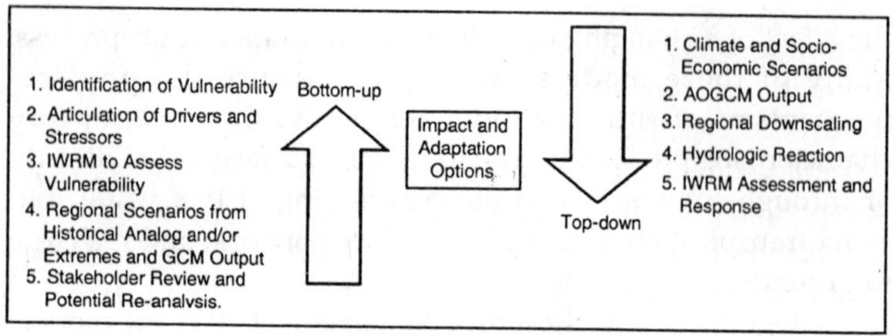

Fig. 1.1 Top-Down vs. Bottom–Up Approaches to Climate Change Assessment.

ADAPTATION TO CLIMATE CHANGE IN THE WATER RESOURCE SECTOR

Water managers have long had to cope with the challenges posed by hydrologic variability, essentially adapting to variability. These adaptations have included the development of coupled reservoir and irrigation systems that allow for the redistribution of water during wet seasons and-normal water years, for carry-over use during dry seasons or to minimize the impacts of drought. Other adaptations, to minimize extensive flooding due to heavy precipitation and high flows, have been the creation of dams and levees to protect cities and concentrated agricultural production.

Protection of growing communities from the risk of flooding has increased in importance in many places. Two adaptation strategies have been the development of flood control operating rules for large reservoirs and the construction of flood bypasses and levees.

These adaptations have had profound effects on ecosystems in many places, and the response has been the establishment of minimum instream flow requirements at important points in watersheds. There is also a growing recognition that ensuring the proper volume of flow for ecosystems is necessary but not sufficient.

Other factors, such as the temperature and quality of the water in rivers, are also important. Recent adaptations with regard to water temperature have included the construction of temperature control devices in large dams that allow for the controlled management of cold and warm water pools which generally develop when large reservoirs stratify. Water quality adaptation measures can include the development of discharge permitting requirements. These have been limited, to date, to point discharges, but are now being contemplated for non-point sources as well.

WATER RESOURCE MANAGEMENT METHOD

The IWRM method is a systematic approach to planning and management that considers a range of supply-side and demand-side processes and actions, and incorporates stakeholder participation in decision processes. It also facilitates adaptive management by continually monitoring and reviewing water resource situations.

To capture the supply and demand side processes and actions, IWRM must simultaneously address the two distinct systems that shape the water management landscape. Factors relating to the biophysical system shape the demand for water (through pricing, incentives for water reclamation and recycling, demand management programmes, etc.), availability of water, and its movement through a watershed; factors relating to the socio-economic management system shape how available water is stored, allocated, regulated and delivered within or across watershed boundaries.

Increasingly, operational objectives of the management system seek to balance water for human use and water for environmental needs. Thus, integrated analysis of the natural and managed systems is arguably the most useful approach.

Groundwater Wells/Groundwater Fluxes

This type of analysis relies on hydrologic modelling tools that simulate physical processes, including precipitation, evapotranspiration, run-off, infiltration and groundwater flow. In managed systems, analysts must also account for the operation of hydraulic structures, such as dams and diversions, as well as institutional factors that govern the allocation of water between competing demands, including consumptive demand for agricultural or urban water supply or non-consumptive demands for hydropower generation or ecosystem protection. Because water quality also will change with climate, special attention must also be paid to water quality changes. Such changes may result in increased restrictions on water withdrawals to maintain water quality and ecosystem health. Changes in each of these elements can influence the ultimate impacts of climate change on water resources.

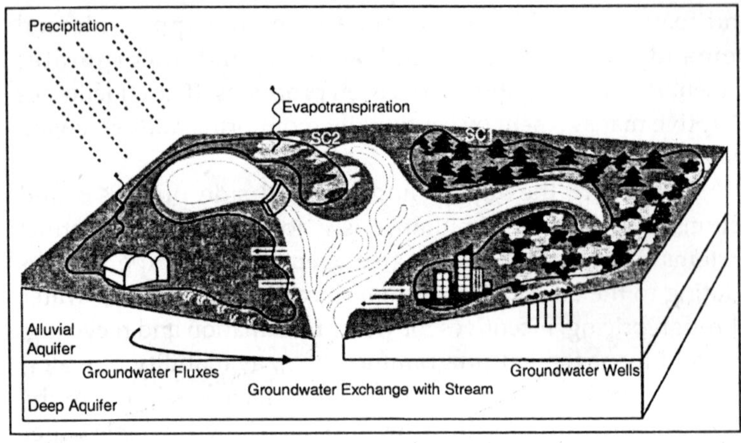

Fig. 1.2 The Interface between the Natural Watershed and Managed System.

Although different hydrologic models can yield different values for stream flow, groundwater recharge, water quality results, etc., their differences have historically been small compared to the uncertainties attributed to global warming reflected in the differences among GCM output.

The chain of effects, however, from climate to hydrologic response, to water resource systems to the actual impacts on water supply, power generation, navigation, water quality, etc., will depend on many factors, each with a different level of uncertainty.

Currently, infrastructure investments and long-term management strategies assume that precipitation and run-off will follow past trends. Mounting evidence for climate change makes this an increasingly tenuous assumption.

INTEGRATED WATER RESOURCE MANAGEMENT MODELS

The water management literature is rich with IWRM models that have focused either on understanding how water flows through a watershed in response to hydrologic events or allocating the water that becomes available in response to those events. Hydrologic simulation attempts to capture the most important land–atmosphere components of the hydrologic cycle. One well-known hydrologic simulation tool is the United States Department of Agriculture's Soil Water Assessment Tool.

The SWAT model has sophisticated physical hydrologic watershed modules that describe, among other things, rainfall–run-off processes, irrigated agriculture processes, and point and non-point watershed dynamics.

The Danish Hydraulic Institute (DHI) offers a suite of models, including MIKE SHE, which is an integrated water resource modelling tool that can simulate all the major processes in the land phase of the hydrologic cycle. The United States Army Corp of Engineers Hydrologic Engineering Center (HEC) developed the HEC-Hydrologic Modeling System (HMS), which simulates the precipitation–run–off processes of dendritic watershed systems. The Delft Hydraulics Laboratory developed the HYMOS model for surface and groundwater hydrology and includes rainfall–run–off simulation.

Important to mention, but perhaps of less importance in terms of water supply and demand management, are hydrodynamics models. These models are typically developed to track the propagation of water through a river system at very short time steps (*e.g.*, minutes to hours). Applications include flood inundation mapping and flood forecasting. From a climate change perspective, this sort of hydrodynamics model could be used to study the interacting effects of sea level rise and changes in freshwater discharge into bay–delta systems at the event scale. Models of this type include Delft's DELFT3D, a sophisticated two- and three-dimensional hydrodynamics model that can simulate non-steady flow and transport resulting from tidal and meteorological forcing. Other hydrodynamics models are DHI's MIKE21 and MIKE3, which can simulate hydraulics, water quality and sediment transport in rivers, lakes and coastal areas, and can be used for local to regional watershed studies. The HEC's hydrodynamics models include HEC–RAS, which is a one-dimensional model for hydraulic calculations and water surface profiles.

This CGE training material focuses more attention on water resource management modelling, which can be used to investigate water supply and demand issues over long planning time horizons, consistent with climate change projections. Water resource management models include the RiverWare DSS, a state–of–the–art water planning and operations model that can be used to develop multiobjective simulations and optimizations of river and reservoir systems, such as and including storage and hydropower reservoirs, river reaches, diversions and water users.

The DHI modelling group provides a GIS-based planning model for studying water management options, referred to as MIKE BASIN. The HEC-ResSim is a reservoir simulation model that can describe operating rules, such as release requirements and constraints, hydropower requirements and multiple reservoir operations, but it too requires prescribed flows from other models.

The MODSIM DSS is a generalized river basin network flow model that can simultaneously incorporate the complex physical, hydrological and institutional/administrative aspects of river basin management, including water rights, but boundary flows must be prescribed.

The MULINO DSS is a multisectoral, integrated and operational decision support system for sustainable use of water resources at the catchment scale, with a focus on the DSS as a multicriteria decision aid. Similar to RiverWare, MULINO can accommodate a physical watershed hydrology model that is external to the system, linked through appropriate input–output procedures. WaterWare is a sophisticated water resource DSS that includes dynamic simulation of physical models of water quality, allocation, rainfall–run–off, groundwater and water management elements, including demand/supply, cost–benefit analysis and multicriteria analysis. Although WaterWare integrates the physical hydrology and the management system, applying the model requires a rather sophisticated level of user and hardware support.

The Delft Hydraulics River Basin Simulation Model (RIBASIM) is a generic water resource planning model for investigating the behaviour of river basins under various hydrologic conditions.

The WEAP21 attempts to address the gap between water management and watershed hydrology, and the requirements that an effective IWRM be useful, easy to use, affordable and readily available to the broad water resource community. WEAP21 integrates a range of physical hydrologic processes with the management of demands and installed infrastructure in a seamless and coherent manner.

It allows for multiple scenario analysis, including alternative climate scenarios and changing anthropogenic stressors, such as land use variations, changes in municipal and industrial demands, alternative operating rules and points of diversion changes. WEAP21's strength is in addressing water planning and resource allocation problems and issues, and, importantly, it is not designed to be a detailed water operations model that might be used to optimize hydropower based on hydrologic forecasts.

The management system in the WEAP21 DSS is described by a user-defined demand priority and supply preference set for each demand site used to construct an optimization routine that allocates available supplies. Demands are defined by the user, but typically include municipal and industrial demand, irrigated portions of subcatchments, and instream flow requirements.

Demand analysis in WEAP that is not covered by the evapotranspiration-based irrigation demand follows a disaggregated, end-use-based approach for determining water requirements at each demand node. Economic, demographic and water use information is used to construct alternative scenarios that examine how total and disaggregated consumption of water evolve over time.

These demand scenarios are computed in WEAP and applied deterministically to a linear-programme-based allocation algorithm.

Demand analysis is the starting point for conducting integrated water planning analysis because all supply and resource calculations in WEAP are driven by the optimization routine that determines the final delivery to each demand node, based on the priorities specified by the user.

USING IWRM FOR ADAPTATION ASSESSMENT

Water managers have long had to cope with the challenges posed by climate and hydrologic variability, both intra–annually and inter-annually. Their adaptation strategies include responding to both seasonal variability and extended wet and drought periods using integrated reservoir and irrigation systems that allow for the capture of water during the wet season for use during dry seasons and extended drought periods.

Other adaptations have been the use of levees and dams in concert to protect communities from heavy precipitation and high flood flows during extreme wet periods. Climate change might challenge these conventional adaptations, forcing more aggressive strategies, such as artificially recharging aquifers or developing integrative strategies that optimally operate reservoirs in conjunction with the artificial recharging of aquifers.

Protecting growing communities from the risk of flooding also has increased in importance. Two adaptation strategies are flood control operating rules for large reservoirs and flood bypasses that move water away from human settlements. In many cases, flood conveyance restricts land use to not allow permanent structures, so during periods of high flow, large volumes of water can be diverted from the main river channels, thereby reducing the risk of flooding along the developed riverfront areas.

Climate change adaptation could include the re–operation of reservoirs to maintain their important services and, in some cases, the construction of new facilities to help in flood protection or to secure water supplies. These adaptations would require careful consideration to ensure their usefulness in adapting to climate change, with an IWRM modelling process a key in assessing the benefits of alternative strategies for adaptation to climate change.

Other water resource adaptations are designed to limit impacts on important aquatic ecosystems. Adaptations include establishing minimum instream flow requirements at important points in the water system. The physical rehabilitation of riverine ecosystems has taken on a priority because the watershed itself can act as a flow regulating system, buffering rapid stream flow response as water slowly migrates through the complex pathways of the watershed.

Water planners now realise that the extensive channelization of many watersheds around the world limits the amount of wetland and riparian habitat available. Setting communities back from the riverine corridor is an adaptation being considered together with the concept that flow-bypass structures can be managed as wetland complexes. There is also growing recognition that ecosystems require more than the minimum flows often prescribed for them.

Other factors, such as the temperature and quality of the water in rivers, are also important. Adaptations with regard to water temperature include the construction of temperature control devices in large dams which allow for the controlled management of cold and warm water pools that generally develop when large reservoirs stratify, which could be amplified with climate change. Water quality adaptation measures can include the development of discharge permitting requirements.

To date these have been limited to point discharges, but they are now being contemplated for non–point sources as well. There has been a clear historical trend towards placing a higher priority on environmental security as societies place greater value on the role of ecosystems.

DATA

Water resource planning models require data on water demand and water supply. Water demand information usually needs to come from local sources–including water use per capita domestic rates, and industrial and commercial water use rates. Common water use rates can be extracted from literature. Irrigation demands can be determined from knowledge about crops and the climate.

Data on demand for cooling water for thermal power plants and mainstream demands for navigation, recreation and hydropower are usually available from the users. Data on ecosystem demand may be available from environmental agencies. All the demands must be adjusted for climate change.

INFORMATION DATA BASE

The water regime in Bulgaria is described over main territorial and temporal units in relation to climate, hydrographic conditions and natural chronological variations. The three main units are: Danube zone; Black sea zone and Aegean zone. The investigation of the river discharge variations in respect to the large territorial units allows to avoid big errors, non-assessment of human impacts over the extended territorial and temporal distributions of river run-off.

The river run-off is assessed as a total run-off of all river estuaries or at country boundaries of Bulgaria for minimal temporal interval of one calendar year. It is assumed that water regulation and transfer from one calendar year to the other using large hydrograph zones do not effect the total annual run-off. A full investigation of the total annual run-off for the 1960–996 period was made. This investigation is used as a base for all other periods with non-sufficient information. Directly and indirectly the data from hydrometeorological stations in the estuaries zones and boundary regions were used: 16 hydrometeorological stations for the Danube basin, 10 stations for Black sea region, and 18-for the Aegean zone.

Time series from representative hydrometeorological stations with the longest period of observation and high–quality data with small human impacts were used to assess the total river run-off. For the period before 1936 the hydrological annual data series selected stations were extended using rating curves. The comparison with precipitation and air temperature was used.

As a result of this investigation the long data series for 16 hydrometeorological stations were obtained. The data series for the precipitation and air temperature were obtained from observations of all rain gauges and meteorological stations. The series were extended using correlative relationships up to 1892. The average values for precipitation (*P*, mm) and air temperatures (*T*,°C) for every year were received using linear averaged interpolation for mean altitude above sea level (*H*, m), in relation to mean altitude of the hydrological zones.

ANALYSIS OF THE MULTIANNUAL VARIATIONS

The objective of the analysis of the chronological variations of the river run-off and the elements of the water balance-precipitation and evaporation, is to obtain the tendency and some possible changes in the future under different scenarios for global climate change. This is the base for reliable estimation of the river run-off and water resources for the last drought period.

Comparison of the tendencies of the chronological variations for precipitation and air temperature with the river run-off gives us possibility to estimate in general the additional evapotranspiration losses in the irrigation systems. In the preliminary analysis of the data for the annual river run-off, the regression and correlation analysis between annual and total chronological values of discharge, precipitation and air temperature for the three main hydrological zones widely were used. As a result some systematic errors from human impacts and information characters were eliminated. Time series were extended up to 1892. We assumed that the hydrometeorological processes are in close relation with the global processes such as solar radiation and activity. Data for the full 10 solar cycles-were used. We obtained the regression values for precipitation, air temperature and river run-off for the years 1890 and 1891 using the long data series for the precipitation over England and Wales and temperature anomalies in the Northern hemisphere. The data were approximated with linear and polynomial (of the power 5) trend curves.

The results show:

1. For all observation period over the sunspots (number of Wolf) since 1700 up to now the increasing linear trend of the solar activity was observed.
2. The same tendency of increasing linear trend was observed for the solar radiation and temperature anomalies over the Northern Hemisphere.
3. The precipitation over England and Wales shows positive trend as well due to the location near Atlantic Ocean coast. Similar correlation between solar radiation and precipitation with time delay for the Pacific Ocean circulation was obtained by Perry, 1992&1993 for the Western coast of North America.
4. The air temperature over Bulgaria since 1890 shows small positive trend in accordance with temperature over the Northern Hemisphere.
5. The precipitation and river run-off for Bulgaria show negative trends; *i.e.* they are in anti-phase with precipitation in England that is determined as a statistical appearance and for atmospheric circulation as well.

In this, we can see that the tendencies for whole Bulgaria concerning air temperature, run-off depth and precipitation are presented.

- Increase in the average annual air temperature for the three main hydrological zones.

- Decrease in the precipitation and river run-off for Danube and Aegean hydrological zones and increase in both for the Black Sea zone.
- The strongest decrease in precipitation and river run-off is obtained for the Danube zone.
- In the Aegean hydrological zone the decrease of precipitation and river run-off is weaker.
- In the Black Sea zone the precipitation and river run-off have small positive trend.
- The main direction in which the humidity trend (P, h) changes for Bulgaria is from West-Northwest to East-Southeast; this fact corresponds with the main direction of the atmospheric circulation and transfer of humid Atlantic air over the country.
- The neutral zone with zero gradients is possibly located around boundary line between the both climatic zones in Bulgaria: European-Continental and Continental–Mediterranean. In the Climatic Atlas of Bulgaria this boundary line is situated along the Black Sea coast and Southwest Bulgaria with main direction from Northeast to Southwest.
- In case of extrapolation in West-Northwest direction from the territory of Bulgaria, such neutral zone with zero gradients obviously may be found, taking into account the anti-phases in the humidity in England.
- The trend values of the humidity obviously can be considered as estimates of the norms (precipitation and run-off) using 50–year period, taking into consideration that the basic period is 107 years and that the straight line of the trend can be defined using two central points from the two half periods (53½ years).

Using as a basis the conclusion, for full century orientation we may calculate the trends values (P,h) for years: 1900, 1950, 2000, 2050 and 2100. The results are presented in Table. Also there the values of the difference P,h=E are presented. These values present losses of atmospheric water for evaporation. The variations of the norms for precipitation, run-off and losses (mainly evaporation) for Bulgaria for the periods of 100 years are shown in Table. The first 100–year period is given as available information, and the second one–as possible value if the tendency in the factors such as solar activity and radiation, atmospheric circulation, human impact will remain the same.

The most important element is the content of green house gases in the atmosphere; different actions are undertaken to decrease it. There is small possibility the decrease in trends of precipitation and run-off to show strong tendency of decreasing for the coming 100-years.

Table. Water Balance Using Trend Estimates for the Two Centuries

Drainage basin	Elements	1900	1950	2000	2050	2100	$\frac{\Delta y}{\Delta P}$	Δy,mm
Total for	P_{BG}, mm	743	728	713	698	683		-60
Bulgaria	h_{BG}, mm	211	198	185	173	160	0.850	-51
	$\alpha=h_{BG}/P_{BG}$	0.283	0.272	0.260	0.247	0.234		
	$E_{BG}=P_{BG}-h_{BG}$	532	530	528	525	523	0.150	-9
	W_h, $10^9 m^3$	23.42	21.98	20.53	19.20	17.76		

The estimates of the norms for the future 100 years using the available trend may be used as a pessimistic scenario for changes in the water resources and the elements of the water balance. The simplest idea is the reverse in the sign of trends for the next century and preservation of the absolute values; *i.e.* the norms for 2050 and 2100 will obtain the values of 1950 and 1900 respectively. This can be used as an optimistic scenario.

DEFINITION OF THE DROUGHT PERIOD

From the analysis is obvious that after 1980 in Bulgaria the continuous decrease in the precipitation with combination of increase in air temperatures appear. This situation leads to depression in the river run-off.

The period in which both run-off and precipitation in Bulgaria are under the norms is 1982–1994 (n=13 years).Before this period the long wet period beginning since 1954 with precipitation and run-off above the norms was observed. During this wet period only during one or two years the values of run-off and precipitation are under the norms.

The chronological structure of the drought period 1982-1994 is executed in relative units–the norms for run-off and precipitation for Bulgaria and for three main hydrological zones are divided to the values of 106–year norms. The year 1995 is the beginning of increase in humidity and probably the beginning of the new wet cycle.

During 1981 the precipitation over the country shows tendency for decrease –99% from the norm, but the run-off is still over the norm with +29%. This delay with one year in the behaviour of the run-off is observed also in the end of the drought period: the precipitation of the 1995 is with 12% over the norm, but the long lasting drought especially in 1993–1994 does not give possibility of the run-off to go above the norm.

The values of lower quartile (25% probability of non–exceedance or 75% probability of exceedance) which may be used as stronger limit of drought are applied.

The river run-off for the Danube basin forms with use of strong criterion one more 13–year competitor drought period

1983–1985. For the run-off in the Aegean basin and for all Bulgaria, the strongest limit gives the 11–year drought period 1985–1995.

This shortening of the drought period from the North to South and especially from West to East, is probably related to availability of spatial waves with anti–phases of the atmospheric circulation in the most important for Bulgaria direction of the humidity transfer from West–Northwest to the East–Southeast.

Similar conclusion was made above during the analysis of trend gradients. Furthermore, we may conclude that the shorter periods with more severe drought (for the run-off) with mainly significance for the whole country, are the periods 1983-1994 (n=12 years) and 1985–1994 (n=10 years).

In the frames of 14-year period 1982–1995 there are two weak raising of precipitation and run-off in some cases: during 1984 –the highest value for the run-off is obtained for the Black sea zone (+44%) and during 1991–the maximal value for the precipitation is received for the Danube zone (+13%).

The probable chronological characteristics of the period 1982-1994 along with estimates of theoretical approximation of empirical distribution curves $P(X < X_P)$ are presented in Table, based on the 106–year period 1890-1995. The theoretical probabilistic curves are lognormal with correction $\ln(x \pm \alpha) = N(X < X_P)$. From Table we may notice that the 13–year period has rather large availability of dry years with low probability of occurrence.

For example, for this short period with probability fewer than 20% only three cases are normally expected. Besides precipitation for the Black Sea zone where such cases are two and for the run-off they are five, for the last zones and for the whole country such cases are from 6 to 8.

Moreover, in this period some years with very low run-off occur having excessively small probability of occurrence-from 0.19 to 0.9%, *i.e.* with return period 526–111 years, surpassing the basic period of approximations. These unfavorable combinations lead to strong decreasing of run-off during the investigated period.

Table. Probabilistic Chronological Structures for the Drought Period P(X<XP),%.

Drainage basin	*	Years												
		82	83	84	85	86	87	88	89	90	91	92	93	94
Danube	*h*	41	20	62	1.7	18	25	23	5.9	2.7	54	34	1.7	0.44
	P	28	26	16	7.8	18	47	38	23	3.1	81	6.3	1.9	16
Drainage basin	*	Years												
		82	83	84	85	86	87	88	89	90	91	92	93	94
Black Sea	*h*	58	26	86	32	31	38	41	15	0.83	28	5.0	11	11
	P	55	34	62	17	21	45	70	36	7.6	74	12	31	25
Aegean	*h*	51	46	65	7.7	17	21	15	7.5	0.9	38	3.1	0.19	0.23
	P	50	39	13	17	25	48	26	21	9.0	37	5.0	0.4	8.0
Total for Bulgaria	*h*	47	32	70	4.6	16	21	17	5.6	0.76	40	8.2	0.44	0.26
	P	41	31	18	12	21	46	36	24	4.3	61	5.0	1.4	11

* Element

COMPARATIVE AND PROBABILISTIC ESTIMATIONS

In Table some 13–year periods are presented chosen from

the 106-year data series concerning run-off and precipitation. In them the minimum values of the average negative deviations are mentioned. It is seen that competitive periods for Bulgaria and for the Danube and Aegean zones are 1982–1994 for precipitation and 1983–1995–for the river run-off (with one year's delay).

Table. Deviation of 13-Year Average Values ($\bar{X}_{13}$) for the Run-off and Precipitation in Relation to Their 106-Year Average Values, ($\bar{X}_{106}$)– $\varepsilon = \left(\frac{\bar{X}_{13}}{\bar{X}_{106}} - 1\right)100\%$ and Return Period N (In Years).

Drainage basin	*	1982—1994		1983—1995		1942—1954			
		ε,%	N, years	ε,%	N, years	ε,%	N, years	ε_m	N_M
Danube	h	–36	3250	–38	4333	–14	65	–38	4333
	P	–14	1560	–12	765	–5	50	–14	1560
Black Sea	h	–28	433	–26	310	–15	38	–28	433
	P	–6	68	–7	44	–9	138	–9	138
Aegean	h	–28	4810	–30	7222	–4	20	–30	7222
	P	–13	1730	–12	1182	–3	30	–13	1730
Total for Bulgaria	h	–31	3333	–32	4063	–8	33	–32	4063
	P	–12	1444	–11	812	–5	48	–12	1444

* Element.

For the Black Sea basin, the minimum value of the run-off is observed during 1982–1994, and for precipitation-1942–1954. In the same Table it is given the estimate for the probable period in which is possible to observe the average value $\bar{\bar{X}}$ (n=13 years).

The sustainable standard deviation of the average value is assessed,

$$-\sigma_{xn} = \sigma_x (N = 106)/\sqrt{n}$$

and for the standardized deviations,

$$f = \left(\bar{\bar{X}}_n - \bar{\bar{X}}_N\right)/\sigma_{Xn}$$

based on Student distribution are received the probable samples of size 13 years $P(t)$, and also the return period,

$$N = n/P(t).$$

From the Table we can conclude that for Bulgaria and for the two of the main hydrological zones the minimum values and respectively the maximum values N_M are observed during the period 1983–1995 for the run-off and 1982-1994 for the precipitation.

In the Black Sea zone values of E_m and N_M are obtained for the run-off-1982–1994, and for the precipitation–the period 1942–1954.

BASIC RESULTS

Since 1980 in Bulgaria long lasting decrease in the precipitation combined with increase in the air temperature were observed, which lead to deep depression in the river run-off. During the period 1982–1994 the run-off and precipitation in Bulgaria are below the norms.

This period is characterized with 31% decrease in run-off for Bulgaria with comparison to the norms to the period 1890–1996 (with deviations–26% from the trend norm and –31% from the average of the basic period 1961–1990).

This period is characterized with very low probability of occurrence: once in 3333, or once in 1444 years for the run-off and precipitation respectively. The relative decreases from the trend in accordance with the absolute value are the biggest for the Danube hydrological basin (– 31%) and the lowest in the Black Sea zone (– 29%).

The return periods are between 1733 years for the Aegean and 722 years for the Danube basins respectively. The drought period 1982–1994 was preceded with the long wet period 1954 – 1981, in which only some years are below the norms. This wet period is included in the study of the 106-year research period with negative trends for the precipitation and the run-off in Bulgaria.

The depressions within the drought period are the most visible during 1990, 1993 and 1994 when the absolute minimums of the longer period–1890-1995 are observed. The deviations are from 0.31 to 0.43 in accordance with the norm and probability of occurrence once in 526 years for the Aegean basin up to once in 120 years for the Black Sea zone.

The drought period of 10-14 years with similar parameters of the discussed one is possible to appear during the next century, although it has very low probability of occurrence.

RECOMMENDATIONS

To decrease the negative impact from the possible future droughts, we can make some recommendations for the utilization of the water resources:

- Building of reservoirs with annual (seasonal) and multi–annual regulation of the run-off as a long lasting prevention action.
- Water transfer from wet water basins to dry regions.
- Rational utilization of the available water accumulated in the reservoirs, natural lakes and ground water reservoirs during the wet periods and years, as a constant issues for the management of the water resources.
- Economical water utilization by all consumers using legislative and economic actions and stimulus.

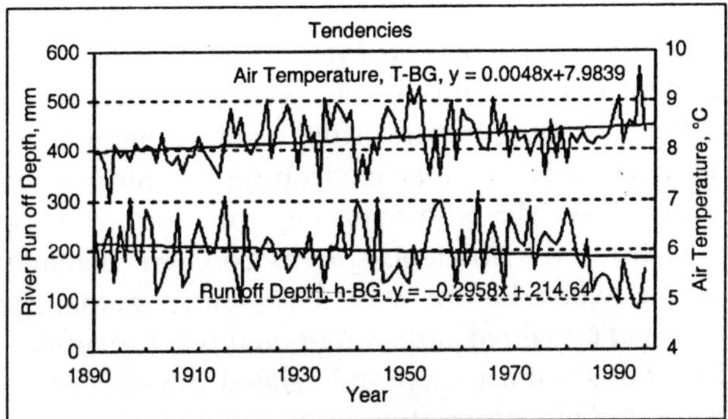

Fig. 1.3 Chronological Graphs of Annual Values for Air Temperature and River Run-off (Depth in mm) for Bulgaria and Respective Linear Trend Lines.

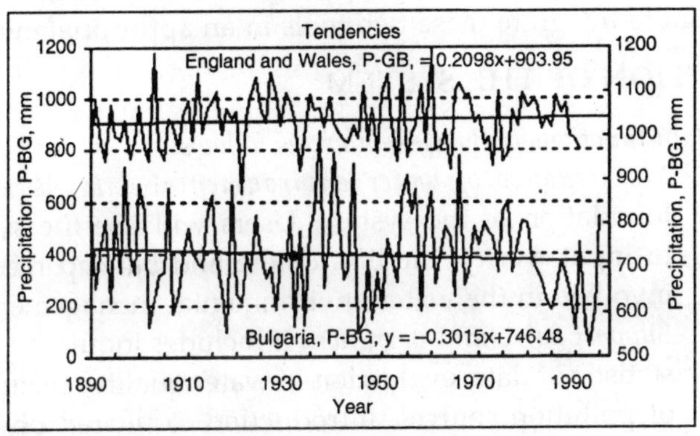

Fig. 1.4 Chronological Graphs of Annual Values for Precipitation Over Bulgaria, Precipitation Over England and Wales and Respective Linear Trend Lines.

WATER ENVIRONMENTAL INFORMATION SYSTEM BASED ON GIS

Water is one of the most important resources of all humanity and plays a crucial role in the development of social economy and people's living. Now the situation of water pollution in China is becoming prominent and more concerned throughout the country. To keep sustainable development, rehabilitation of polluted water bodies is a task should be accomplished by great effort. Water Environmental Management Information System (WEMIS) is established to achieve the collection, transmission, storage, maintenance and analysis of all sorts of water environmental information. The construction of WEMIS base on new information technology can greatly upgrade the efficiency of environment management. Information involved in the water environmental management can be divided into three parts: water quality information, water pollution source information and related physical geography and socioeconomic information.

With the support of modern information technology such as GIS, WEMIS can rapidly employ the information well and truly. GIS and water environmental models are two analytical tools to establish a decision support system. It is necessary and potential to integrate these two tools in an appropriate way.

FUNCTION OF THE SYSTEM

Main functions of the system are as follows:

- *Maintenance of water environmental data*: It is the foundation of the system. Users will use the system to input, verify, modify, delete and backup the data involved in the water environmental management.
- *Support for routine operation*: It includes inquiry of data, statistic of data, evaluation of water quality, evaluation of pollution sources, introduction to related physical geography, socioeconomic information and etc. Uses can comprehend the water environmental situation visually and quickly by all kinds of tables and graphs.
- *Decision support for management*: It includes prediction of pollution sources, prediction of water quality, calculation of water environment capacity, allocation of pollution load and etc.

STRUCTURE OF THE SYSTEM

The system comprises six sub-system, and data flow diagram of the system is Figure1.5?

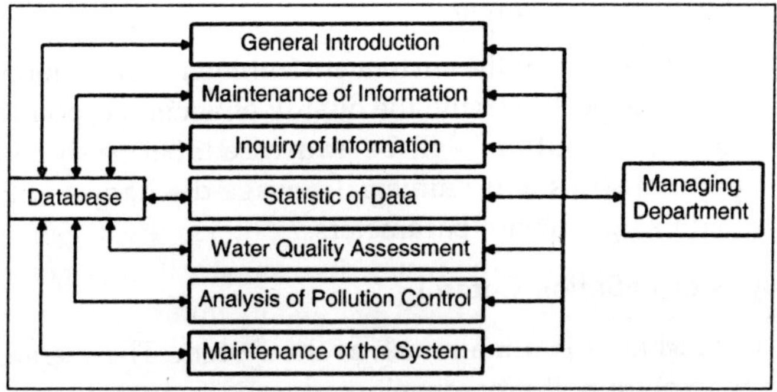

Fig. 1.5 Data Flow Diagram of the System

General Introduction

In this part, geography and socioeconomic situation, development of water quality and pollution sources, distribution of investigation station of a specific area can be comprehended in various formats such as text, table, picture and video etc.

Maintenance of Information

It includes input, modification and deletion of all the related data. In order to keep the integrality and consistency of the data, the system can provide validity verification.

Inquiry of Information

The system can form the corresponding Transact-SQL commands according to the demand of the user and send the request from the client application. Then the commands will be disposed by DBMS on server and return results to the client application.

Statistic of Data

Data statistic is an important working content in water environmental management. The system can conduct various statistics and generate report forms. Major items include monitoring station, water quality, and water pollution source.

Water Quality Assessment

The assessment of the current environmental situation will focus on two aspects. Firstly, the quality of water, *i.e.* pollution levels and distribution, will be addressed. Secondly, main polluting industries and municipal sewage discharges will be expressed using relevant parameters.

Analysis of Pollution Control

Data Mining is a main task of the system. The managers need to explore pollution-developing trend from the common transaction information and find necessary decision for pollution control. A series of water environmental models are involved in the system. These models include 1D and 2D water quality model, prediction model of polluting source and optimum method of polluting load allotment.

MAINTENANCE OF THE SYSTEM

In order to guarantee the security of the information of the system, all users will be assigned to a characteristic that determines the operations each user can perform.

INTEGRATION OF GIS WITH WATER ENVIRONMENTAL MODELS

GIS and water environmental models are two main analytical tools applied in the system to support decision-making. GIS based on common application provides measures for managing spatial information, but it can only act as system software.

For applying all functions of GIS to the water environmental information system, GIS should be integrated with application models in the corresponding field. One the other hand, water environmental models have distinct spatial character. Water environment situation is expressed by parameters indicated by monitoring cross-section and monitoring spot. The models can distinctly reveal the water environment quality in different area by introducing GIS, and indicate the distributing tendency of water environment quality.

APPLIED AREA OF GIS

GIS can help water environmental models in the aspect of data preprocess and data acquisition. It can also be used to establish postprocessor to output the result of water environmental models visually. For 1D water quality simulation, it can be efficiently applied to spatial display and analysis. With interpolation and contouring it can indicate the 2D water quality dispersion.

Spatial inquiry can be performed with GIS associated with result of water environmental models. For example, by using water quality model and line and polygon overly analysis technology in GIS, we can have comprehensive forecast and assessment over multiform pollution index of one water regime.

So integration of GIS with water environmental models is needed and potential. An ideal structure of integration should be adopted in the water environmental information system.

INTEGRATION METHOD

Different methods of integration differ from each other in two aspects: data interchange and user interface.

In the following discussion, four methods of integration will be analysed:

1. Models and GIS are separate. They have user interface of each own, and exchange data with file. In this way the user interface won't have uniform style and the efficiency of data exchange is low.

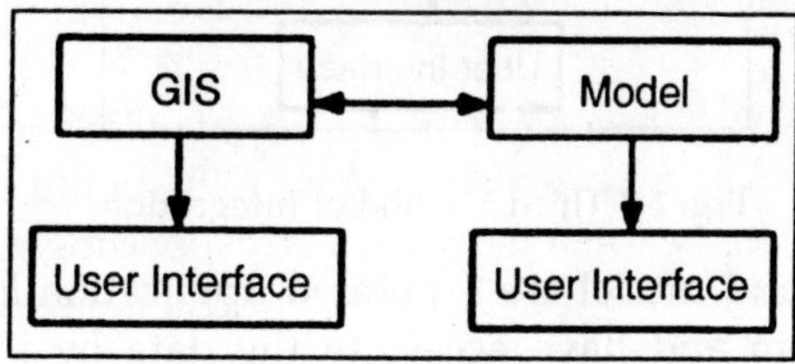

Fig. 1.6 First Method of Integration

2. Models become an inline programme of GIS?*i.e.* Model is one module of GIS. Indeed some GIS products have

analytical models can be applied in the water environment, but it is impossible expect one product can solve all the problems.

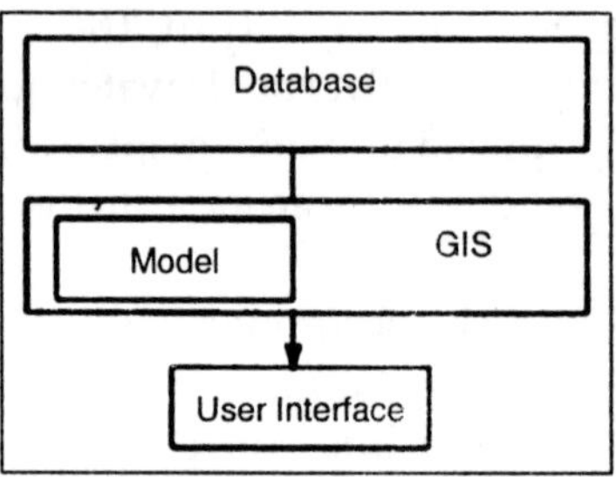

Fig. 1.7 Second Method of Integration

3. GIS becomes an inline programme of model *i.e.* Models have some function of GIS. In the past, some researchers developed postprocessor to visualize the result of model. Although it improve the visual effect of the simulation, it fairly difficult to achieve the spatial analysis function of professional GIS product.

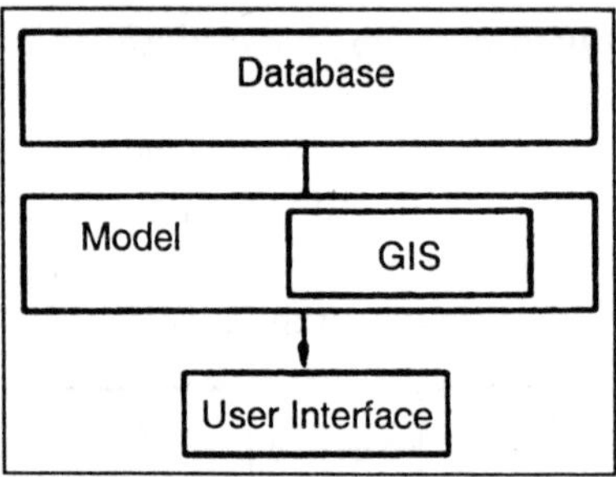

Fig. 1.8 Third Method of Integration

4. GIS and model are separated application level of the system and have access to the data by information sharing. Users can operate GIS and model through a common interface.

   In this way, GIS and models can be integrated in a seamless method. Some calculating conditions of the model will come directly from database and spatial analytical function of GIS can be involved in the model. On the other hand, it becomes easier for the developers to establish a system that is more efficient.

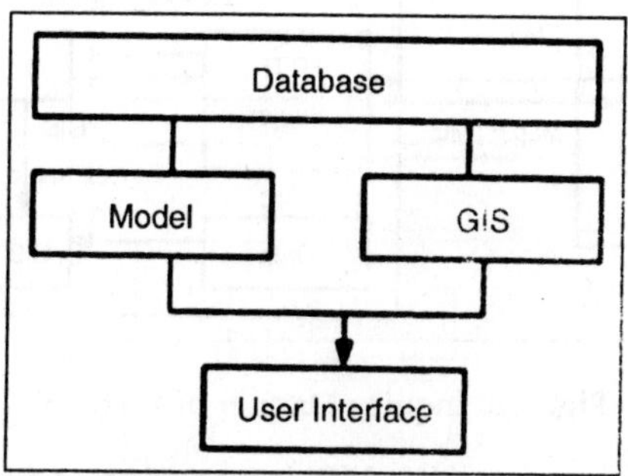

Fig. 1.9 Fourth Method of Integration

APPLICATION OF WEBGIS

In recent years Internet is rapidly spreading throughout the world. The Browser/Server (B/S) structure are more adopted by all kinds of information system. Contrasting to Client/Server (C/S), B/S has more flexibility and expandability and is easier to maintain, moreover, it can make the environmental information more open to the public.

But we aren't mean to substitute B/S structure for C/S structure. Instead, we will use both of them in the WEMIS. Information maintenance and analysis of pollution control will adopt C/S, and those functions that is more tied with information inquiry will adopt B/S.

WebGIS should be taken in the information system of water environment. Users will use Internet browser programme to send requests that can be accepted by Web server. The Web server startup CGI, ISAPI or NSAPI programmes and transfers the requests to GIS system. Accessing data by ODBC, GIS system will process the requests initiated by a client and return result.

High-speed data transfer and high-performance network server should support WebGIS. But it won't be obstacle anymore in the near future with the fast development of modern network technology.

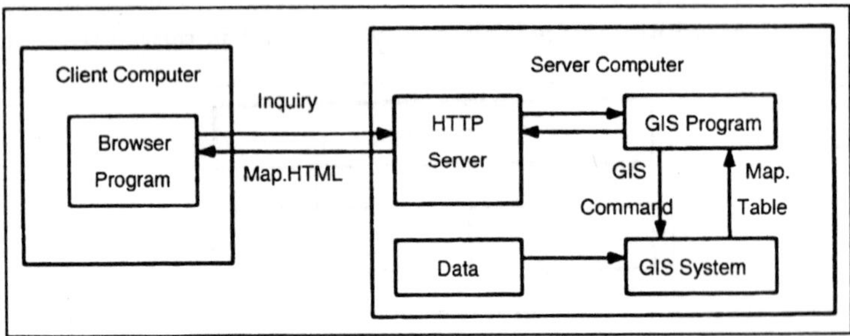

Fig. 1.10 Inquiry Manner of WebGIS

The development of WEMIS is an active and potential area. New technology such as Data Warehouse will be introduced into it to enhance the decision support capacity.

In the construction of WEMIS, GIS is an important technical element. The application of WebGIS is now at an elementary stage. There is a continuing need for more comprehensive research and practice on the application

SIMPLE ANALYSIS ON FUTURE WATER RESOURCE

Recently many researchers have paid attention for water resource problems and many people are worrying about water conflict in the future. World water forum indicates probability of water shortage by increase of irrigated area with population increase and said right of secure water use 40 liter per day for all people in all over the world as target. However the achievement of this target is not only difficult but also lack of water happens everywhere.

Such problems are complicated in an international river and international organizations are needed for management of the river. The Mekong River is going through six countries and the river has 800,000km2 as the area with 4,200km length. Lack of water is worried in the future because Lao and Cambodia are developing for irrigation and hydro power plant. Especially Vietnam makes caution for such water resource developments.

The design of development in the region needs holistic and comprehensive planning and the evaluation and the assessments. Especially, multi– directional and multi-scale discussions are necessary for development in the headstream of the river.

MRC (Mekong River Commission) compiles database consisted of hydrologic and meteorological data in lower four countries. Also remote sensing and digital map data are available and useful supported by GIS. The ultimate target is discussion on the development design by multi-direction and -scale organically using those data. In this paper, simple analysis of water balance and drought problems are mentioned.

DATASETS

Satellite data of AVHRR/NOAA is used for grasping vegetation condition and land use. The data is received and complied in 1998 by ACRoS (Asian Center of Remote Sensing) in AIT(Asian Institute of Technology). Meteorological and hydrological data are obtained from MRC Year book in 1993. Elevation data is obtained from GTOP30 of USGS and extracted as watershed area. Land use data is depending on IDI (Infrastructure Development Institute, Japan). Study basin is considered only lower area from Chanseing.

DROUGHT PROBLEMS WITH LAND USE CHANGE

Study Basin

Virtual area was introduced in this study for simple consideration without complex land use and reservoir management. The virtual area is given as unit area with only three land uses, which are forest, paddy and urban area. Considering the Mekong basin, upper and lower regions are covered with forest and paddy field, respectively. Urban area is less than 1%. It was supposed that urban area is developing in the lower region. Land use data in 1997 show us 17% crops and paddy fields. Other areas are forest and glass lands. According to development, crops area will expand from flatter area. Potential area for crops is obtained from slope calculated elevation data. Ratio of slope less than 20m/km to whole area is 58%. Now supposed rate of land use change on time is supposed. Crops field linearly increases until potential area and simultaneously forest area decreases. Urban area increases in crops field until 10% in the same time range.

Hydrological Processes

Precipitation data at Pakse is used as input to study basin and is distributed uniformly in a whole basin. It is supposed that evapotranspiration in crops field was estimated as potential evapotranspiration 120mm/month in period of water supply. Kazama etc. estimated monthly evapotranspiration from forest in Thailand and Vietnam.

Those values are used. In urban area, only interceptional evapotranspiration is considered as 10mm/month, where monthly rainfall is equaled to urban evapotranspiration if the rainfall is less than 10mm/month.

Although crops field area is grater than paddy field now, it is supposed that paddy field will dominate crops field by irrigation system expansion.

The storage height of paddy field is 100mm/month from May to September, rainy season. When the result of water balance is not enough for the storage, the lack will be charged by run-off after June. Storage period is from May to September and all storage water discharge in October. Storage in urban area is ignored. In forest area, run-off and storage is expressed by the Tank model.

Run-off ratio to net rainfall is 0.9 and infiltration ratio coefficient to storage height is 0.02. Concerning infiltration in paddy field, the rates are 2mm during storage period and 2% to net rainfall during unstorage period, respectively.

When run-off is negative by water balance, paddy field releases storage water to river, which has 5mm/month as minimum discharge. After this the supplement for paddy field is charged from run-off of next month. These conditions are shown in Table.

Table. Input Hydrological Amount for Land Uses

	Forest	Paddy	Urban
Rainfall P	Constant: Pakse: MRC Hydrological Year Book, 1993.		
Runoff R	Calculation by Tank	Water balance	
Infiltration I		Dry: 2mm Wet: 0mm	0mm
Storage S		Full: 100mm Not Full: 0mm	0mm
Evapotranspiratione	Table	Full: 120mm Not Full: Forest.	Interception: 10mm

Table. Evapotranspiration (mm)

Months	Paakse	Tokyo	Months	Paakse	Tokyo
Jan	10	10	Jul.	140	120
Feb.	10	10	Aug	150	120
Mar.	20	30	Sep.	150	90
Apr.	30	60	Oct.	100	50
May	100	80	Nov.	10	30
Jun.	130	100	Dec.	10	10

Estimation of Drought by Water Balance

Water balance is calculated by the following equation.

$$R = P + Q - E - I - \Delta S$$

Where R is run-off, P is precipitation, Q is discharge, I is infiltration and deltaS is storage. Every amount is used as mentioned. Run-off in the study basin is estimated from water balance considering the land use change in time. Figure 1.10 indicates water balance in cases of time 0 and 80. Here, time is conceptual under virtual scale. Although there is a little bit of difference, evapotanspiration in summer and run-off in June and July are decreasing in time 80 with deforestation. The reason is that storage water is increasing as crops area is increasing. In summer period, evapotranspiration in forest also is more and run-off is increasing, on the other hand, less run-off produce. Lack of water happens during planting season and it will influence to the month of May after the expansion of crops area. To compare with these results, the calculation of precipitation and evapotranspiration in Tokyo in the same land use as the Mekong. This will explain the importance of availability in water resource in planting season on the scenario, which is expansion of paddy field. In Japan, snowmelt water can supply water for paddy field in this season, however the Mekong region has drought. The existence of dry season makes chronic drought in the Mekong. Water demand is also increased not only by expansion of crops but also urban area. Urban water demand height is calculated as follows.

$$Wd=U_w \times Pd$$

Where Wd is household water height, Uw is urban demand, 20m^3/month and Pd is urban population density, 6,000people/km^2. Urban water demand is not variable in Japan in recent years, so that, the value in Japan is useful. Urban population density is referred from Singapore and Hongkong. As the result, Wd is 120mm/month.

Drought happens when water demand excesses run-off from the basin. This means R<Wd at lowest point. The number of drought months by land use change is indicated. This shows Tokyo's phenomena. In the Mekong region, the number of drought months increases as crops area increases. The reason is why base flow from forest decreases and storage amount increases. In time 0, drought happens almost once a month, which is season of plantation. Widening paddy field extends drought months after May. In the case of Tokyo, decrease of drought months is shown due to decrease of evapotranspiration in urban area. These results have some assumptions and difference from actual phenomena.

RELATIONSHIP BETWEEN POPULATION AND IRRIGATION AREA

The importance in water resource management is practice of water supply for everyone as mentioned in introduction. Therefore, let us consider potential irrigation area to distribute 40l/month to every person as target of World water Forum. Monthly water demand Dw is,

$$Dw = Pd \times Dw_p$$

where Pd is population density and Dwp is personal water demand with 40l. Population density 85 people/km^2 is estimated from 67.4million people and 793.1 thousand km^2 in the Mekong river basin. Using these values, Dw is 0.102mm/month. In February, which is the most drought month, run-off from study basin is 15mm/month obtained by water balance. In this case, evapotranspiration in forest is equal to crops field. If 100mm of water is necessary for irrigation in dry season, irrigated area Ai is,

$$Ai = (R-Dw)/100$$

where R is 15mm and Dw is used as above results. The variation of potential irrigated area is if population will be double. The Mekong river basin shows a little chance of population increase. Almost current crops field can supply to double population, however, little expansion of irrigation area rapidly decreases water supply water for one person. In this result, 15% irrigation area to a whole basin is maximum and the expansion more than this needs more efficient utilization for water. Expansion of irrigation area is national policy in Laos and Cambodia and it will be apprehensive about water supply in the future.

SUMMARY

Water resources in the Mekong basin are quantitively evaluated from water balance calculation using the approximation of simple land use change.

In spite of many approximations, it shows that some interesting results, which are as follows:

- Wide area for potential irrigation in lower region,

- High influences between irrigated area and run-off amount,
- Increasing number of drought months by expansion of irrigation area, and
- Apprehension of water lack in the future.

2

Water Resources Management

INTRODUCTION

Siberia has the largest area and the richest natural resources in Russia. Hydrographycally, a larger part of West Siberia coincides with the basin of the Ob-Irtysh River system. Among the rivers falling into the Arctic Ocean, Ob is the longest river (together with Irtysh) with the largest catchment area. Siberia covers more than $10 \cdot 10^6$ km^2. West Siberia occupies 2 500 000 km^2. The common characteristic of the Siberian geographical structure is its hydrographic unity and unique hydrologic regime of rivers. The overwhelming majority of rivers belong to the Arctic Ocean basin, and only a small part of the Transbaikalian region is drained by tributaries of the Amur River. Ob, Yenisei and Lena, which are among the world's largest rivers, constitute the basis of the river network in Siberia.

Most of the Siberian rivers are fed by thawed water in spring and by rains in summer and autumn. Ground feeding is usually low, often below 10%. For almost all the rivers, 80–90% of annual flows occurs during warm periods and no more than 7-15% in winter. A great part of Siberia is covered by wetlands and bogs. The area of wetlands in the West Siberia is estimated as $800 \cdot 10^3$ to $900 \cdot 10^3$km^2 including about $400 \cdot 10^3$km^2 of peat bogs.

THE OB-IRTYSH RIVER BASIN

Based on water run-off, Ob is the third largest river in Russia. The river is formed by the confluence of mountain torrents Biya and Katun, both originating in the Altai Mountains. Ob drains an area of $2.99 \cdot 10^6$ km^2 (including internal closed drainage area of $0.53 \cdot 10^6$ km^2) and extends 3,650km in length (from the source of Irtysh 5,410km). Mean annual run-off is 402km^3. The bottom is mostly sandy, with rapids. Ob is a typical river of the plains, its mean gradient being 0.000042. The main tributary of Ob is Irtysh which takes its source south–west of Altai Mountains (Mongolian Altai) in China. The total length of Irtysh is 4,248km, the watershed area is estimated as $1.6 \cdot 10^6$ km^2. Mean annual run-off is about 90km^3.

Intensive development of various branches of economy in Siberia gave rise to many complicated environmental issues relating to water resources management and protection as well as to use and disturbance of natural systems including those closely and inherently connected with hydrologic systems.

THE UPPER OB RIVER AND THE NOVOSIBIRSK HYDROPOWER RESERVOIR

Unlike the Irtysh, Angara and Yenisei Rivers, Ob itself has only one hydropower station. That is the Novosibirsk HP Station (455MW) with a rather large and relatively shallow reservoir (volume of 8.8km^3, surface area of 1070km^2, average depth of about 9m). The station was built and the reservoir was filled by 1959. The effect of the impoundment on river sediment transport resulted in a degradation of river channel downstream of the dam. This adverse effect was aggravated by dredging of alluvial material from the river channel for construction works near Novosibirsk.

The river bottom and, as a result, water level dropped up to 1,5-2,0m and that brought about serious difficulties in water supply (associated with withdrawing water from the river), in particular during low–water periods. Other environmental impacts include considerable abrasion of reservoir shores and impounding of groundwater near shorelines.

THE ENVIRONMENTAL IMPACT OF OIL AND GAS INDUSTRIES IN THE MIDDLE OB RIVER BASIN

Ecological well-being of the substantial part of Ob River basin is heavily affected by the chemical contamination of hydrologic systems, including surface and ground waters. The industrial development in Siberia gave rise to various kinds of environmental damage and sometimes–even devastation. The damage includes distortions in the behaviour of hydrologic and hydrogeologic systems and their qualitative state.

There are many examples of large-scale local contamination of lands and waters in Siberia, because of various industrial activities, such as oil and gas production, processing of mineral, wood and other natural resources and use of transport systems. Oil refineries and metallurgical plants contribute to both direct and indirect contamination of watersheds and hydrologic systems due to discharge of waste waters, emission of gaseous effluents and disposal of wastes in tailing dumps.

Large-scale pollution of waters and soils occurs in the Ob-Irtysh River too. Some small and mid-size tributaries of these large rivers, such as Inya, Tom, Chulym, Vakh, Ishim and Tobol are heavily contaminated. Previously, river pollution occurred mainly in the upper stretches of Ob and Irtysh. In the last thirty years, however, industrial and urban pollution involves larger areas reaching the northern regions. That is the negative result of the development of West-Siberian oil and gas industry.

Therefore, the problem of oil contamination of watersheds, rivers and their flood plains in the Middle Ob River basin has become most urgent today. Accidents at oil pipelines, wells and pump stations are the main cause of contamination by hydrocarbons and salt (brine). The use of slime storage pits at well sites and release of oily wastes from technological systems also play an essential part in the contamination of hydrologic systems. Small rivers and some tributaries of Ob situated in or near oilfields are subjected to contamination in the most considerable degree. All this results in the large-scale and substantial hydrocarbon contamination of soils and waters in the Middle Ob River basin. The salinization of river waters by brine amplifies the environmental impact. Chemical contamination of rivers in the North Tyumen area and the rivers' low ability for self–purification already brought about negative consequences for aquatic ecosystems. The most visible of them is the damage done to fisheries in the Middle and Low Ob River.

ENVIRONMENTAL IMPACT OF HYDROPOWER DEVELOPMENT IN THE IRTYSH RIVER BASIN

Three hydropower stations were built on the Irtysh River: Bukhtarminskaya, Ust-Kamenogorskaya and Shulbinskaya. The biggest of them is Bukhtarminskaya HP (675MW) with a large and deep reservoir (volume of 49,7km^3, max. depth of about 70m). The station was completed by 1966 but the filling of reservoir with over–year storage turned out not an easy problem. The main adverse consequence of the project was the substantial change in hydrologic regime of the river downstream of the reservoir and the resulting impact on water regime of flood plains leading to their aridization. In the Omsk Region alone, about 300,000 hectares of floodplain were damaged, including 200,000 hectares of agricultural lands. About 30% of such land dried up.

THE TOM RIVER PROBLEM

The Kuzbass Industrial Complex which is the most important center of coal mining in Russia and one of the centers of metallurgical and chemical industries gives another example of a large-scale environmental impact. This area is located in the Tom River basin (the right tributary of Ob) and includes such industrial cities as Kemerovo and Novokuznetsk. In this case, the complicated and tense environmental situation resulted from the fast development of a number of industrial activities, including coal mining and processing of ferrous and non-ferrous ores. Together with the growth of urban settlements, industrial activities and lack of environmental measures has caused substantial pollution of air, lands and waters. The Kemerovo Region is the most densely populated and heavily industrialized area in Siberia with supporting agricultural activities. In that region, some urban areas, industrial sites and stretches of the Tom River and its tributaries are badly contaminated.

The Tom River is 827km long and drains a watershed area of 62,000km^2. The total mean annual run-off of the river is 34.1km3/year. The basin contains several environmentally devastated urban and industrial areas of the Kuzbass Industrial Complex. The Tom River system is the main source of water supply for the Kuzbass region and the ultimate collector for industrial, municipal and agricultural waste waters. Over a thousand industrial plants and human settlements discharge inadequately treated waste waters into the Tom River and its tributaries. Washout of pollutants from municipal and industrialized areas and pollutant run-off from farmlands are also very substantial. As a result, the river waters contain practically all types of pollutants.

The hydrologic regime of the river features a very nonuniform seasonal run-off distribution. This reflects its mixed origins with prevailing input of snow melting (about 70% of the total run-off is released during spring flood period). Winter discharges are very low (due to limited groundwater feeding); the part of the total run-off related to this period is about only 5%. Such hydrological conditions are most unfavorable both for water supply (lack of water and low levels) and for the quality of river water. The latter results from the fact that the concentrations of a number of point-source pollutants reach high levels because of low dilution.

A project has been proposed in the 1970s aimed to mitigate the water quality state and the hydrological regime along the lower part of the river: near and downstream of Kemerovo. The concept of the project was to create a rather large river reservoir by constructing a dam near Krapivino, a settlement upstream of Kemerovo, and to regulate river flow providing sufficient discharges to dilute the pollutant load during the low-water periods. Construction works started in 1975 and were conducted at a rather slow pace because of financial constraints. In the late 1980', the project became a subject of criticism by some public groups and politicians reflecting the emergence of the "green" movement in Russia at the time. During the dramatic political events in Kuzbass (the strike of miners) in 1989, the construction operations at the dam were suspended to meet the demands of the striking committee provoked by the "green" groups. According to a governmental decision, the project had to be considered once more with an additional environmental impact assessment. Based on the analysis of environmental situation and the consequences of interrupted construction works, after long public discussions, the Administration of the Kemerovo Region recently came to the conclusion to review the project design with a new EIA and then, possibly, complete construction works.

From 1990 through 1992 (and now again), the Institute for Water and Environmental Problems and other institutes of the Siberian Branch of the Russian Academy of Sciences carried out a research programme whose main objective was to assess the water management and ecological state of the Tom River and its basin as well as the engineering measures proposed to improve water resources and quality management.

This wide–range multidiscipline study included collection and analysis of available hydrological, hydrochemical, ecological and other data on the river system and its watershed area, field investigations and the environmental impact assessment of the Krapivino project.

The field investigations pursued mainly the extensive survey of the necessary hydrochemical, biogeochemical and hydrobiological data on the state of the Tom River basin, including quality of surface and ground waters; pattern of sediment, soil and snow pollution; species diversity under the anthropogenic stress and bioaccumulation of chemicals by the river inhabitants. The hydrochemical study of water quality reveals that the Tom River and some of its tributaries are heavily exposed to anthropogenic contamination especially immediately downstream from the large industrial centers. Main pollutants include organic compounds (petroleum compounds, phenols, polycyclic aromatic hydrocarbons, formaldehyde, aniline, organic chlorine compounds, some amines, naphthalene and its derivatives, dibutylphtalate and its derivatives), nitrate and ammonia nitrogen as well as some heavy metals (cadmium, zinc, chromium, copper, etc.). Concentrations of the above substances often exceed drastically the national standards for water quality in natural water bodies.

Presently, the quality of surface waters in the Tom River basin is determined to a great extent by pollution from different sources, including industrial plants, urban sewerage systems, agricultural lands and urban and industrial areas. Therefore, its essential improvement is impossible without significant reduction of quantities of contaminants emission to air and of water and soil pollution from industrial and urban sources, as well as more careful use of fertilizers, pesticides and other agrochemicals. Esearch aimed at establishing the scientific basis to develop a strategy for improving the environmental state of the Tom River and its basin. The results obtained thus far have demonstrated conclusively that the water quality problem can not be considered separately from other environmental issues of the region. It can be done only within the total environmental context of the river basin. There is a need for developing a strategy of integrated management of water and environmental quality on the river-basin scale. As to the fate of the Krapivino reservoir project, the final decision regarding the continuation of construction works is yet to be made.

The Tom River problems and the situation with the Krapivino Reservoir works represent quite well the character of many problems in Siberia and, generally, in Russia, which lie on the interface between water resources management and environmental protection.

RESEARCH ON ALTITUDE DISTRIBUTION OF SNOW DEPTH IN WIDE AREA

INTRODUCTION

The winter monsoon from Japan Sea brings heavy snow to East Japan, which depth is sometimes over 5m. Therefore, east Japan is famous for one of the heaviest snow area all over the world. Heavy snow causes traffic jam, cutting electric wires, prevention of commercial transportation and disasters such as house collapsing, snow avalanche and snowmelt floods. On the other hand, snowfall in winter is stored in the mountains and becomes useful as water resources for electric power, irrigation and water supply.

Therefore, estimating snow volume in a wide area is important for water resources planning. However, the characteristics of snow distribution have not been understood clearly because of difficulty of snow survey at high elevation region and difference of annual amount of snowfall. Many researchers reported that snow depth is related to elevation. Koike estimated snow water equivalence from satellite images. Lu retrieved regional character of snow distribution of snow distribution with snowfall simulation.

These study areas are limited only one watersheds. Kazama et al estimated snow distribution in the Tohoku district, Japan. But this research supposed snow depth is proportion to the difference of elevation from snow line and detailed character of snow distribution cannot be estimated. Matsuyama arranged snow surveys conducted in mountain regions. Every study periods of these snow surveys vary according to snow survey.

Fig. 2.1 Snow Map

The purpose of this study is to understand the relationships between snow depth and altitude distribution in wide area (250,000km^2).

STUDY AREA AND DATASET

East Japan (250,000km^2) was selected as a study area. It is located between 35°N and 45°N. This area includes Hokkaido Island, Tohoku district, Hokuriku district and the Japanese Alps which are heavy snow area in Japan. Winter monsoon brings high humidity from the Japan Sea to the mountains and results heavy snowfall which depth exceeds over 5m. Almost all regions are steep and covered with the natural dense forest.

There are about 200 points AMeDAS (Automated Meteorological Data Acquisition System, Japan Meteorological Agency) snow depth observation stations at study area. Those data were used as ground truth data of snow depth. Geographical Survey Institute, Japan publishes various kinds of digital maps. Ks–110 among them is used as DEM (digital elevation data). One Snow covered area (Snow map) images per one month were prepared. These are composed with NOAA/AVHRR satellite data and multi- spectral analysis. Figure 2.1 shows an example of snow map.

THE EVALUATION OF ALTITUDE DISTRIBUTION OF SNOW DEPTH

The Distribution of the Gradients of Snow Depth to Elevation

As the index to evaluate the altitude distribution of snow depth, snow increasing coefficient is utilized. This is the gradients of snow depth to elevation and is presented following,

$$\alpha = \frac{SD}{\Delta H}$$

where, a is snow increasing coefficient (cm/m), SD is snow depth (cm) at AMeDAS snow observation point and ΔH is the difference of elevation from snow line. Figure2.2 show the distributions of snow increasing coefficients at AMeDAS snow depth observation points. The values of snow increasing coefficients at the region near the snow line are very high. Specially, these are very high in the low elevation area where snow depth is high because the difference of elevation from snow line is very small. On the other hand, the values of coefficient are low in high mountain area. Main Island of Japan can divided into Japan Sea side from Ohu mountains range and Pacific Ocean side from this. Winter monsoon from Japan Sea brings to Japan Sea side. Low elevation area in Japan Sea side is covered with snow. The difference of elevation from snow line is large and snow increasing coefficient is low. Comprising this area, in Pacific Ocean side is not affected by winter monsoon. Low elevation area is not covered with snow, therefore the difference of elevation from snow line is small and snow increasing coefficient is high.

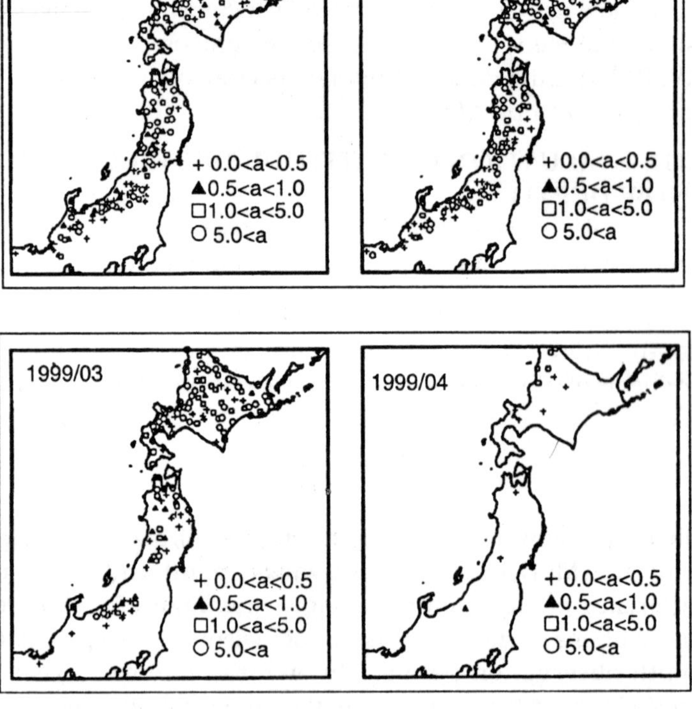

Fig. 2.2 Distribution of Snow Increasing Coefficient

Figure 2.3 shows temporal variation of snow increasing coefficients under 5.0. The abscissa indicates cumulative day from 1st December, 1998. In many regions, maximum coefficients are appeared at February or March.

This reason is that maximum snow depth is appeared at February and March. The changes of snow increasing coefficient of which maximum is under 1.0 are stable. These are in high elevation area of which the condition of snowfall is stable. On the contrary, the others is not stable. These are the one in low elevation region. These depend on condition of snowfall.

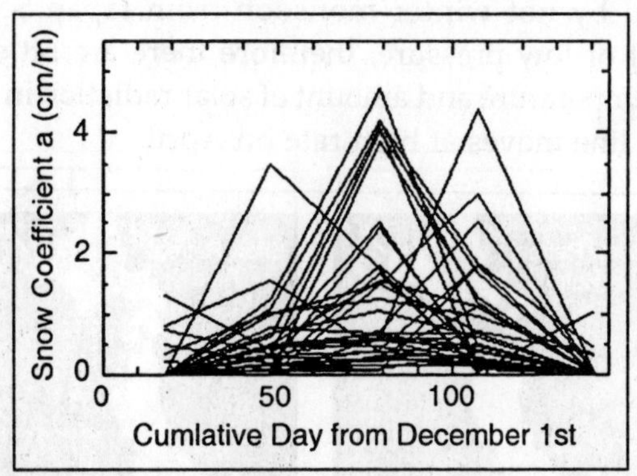

Fig. 2.3 Temporal Variation of Snow Increasing Coefficient

Retreat Rate of Snow Line

Snow line retreats from low elevation area to high elevation area in snow melt season. The retreat of snow line influences the altitude distribution of snow depth. Therefore, the distribution of retreat rates of snow line is estimated from February to March and from March to April 1999. The method of estimating retreat rate is following. Comparing snow maps of current month and previous month, if the area, where previous month is covered with snow, is not covered with snow, the distance from nearest snow line at previous month, in short retreat rate as one month, is calculated.

Figure 2.4 shows the images of distribution of retreat rate at March and February 1999. Noticeable retreats of snow line are appeared at March. These also make sure snow map on February and March. This snow season is little snowfall. Kazama[5]) reported that low elevation areas are not covered with snow in little snowfall year, on the contrary, these areas are covered with snow in heavy snow year. This snow season is little snowfall and low elevation areas are not so covered with snow, therefore snow lines don't retreat. The areas where snow line moves are valley and near the coastline. Retreat of snow line appeared on April. Specially, southwest at Hokkaido Island and Kitakami mountain range is notable. These area are influenced by not winter monsoon from Japan Sea but the movement of low pressure, therefore there are little snowfall and low temperature and amount of solar radiation in these area and snow line moves at high rate on April.

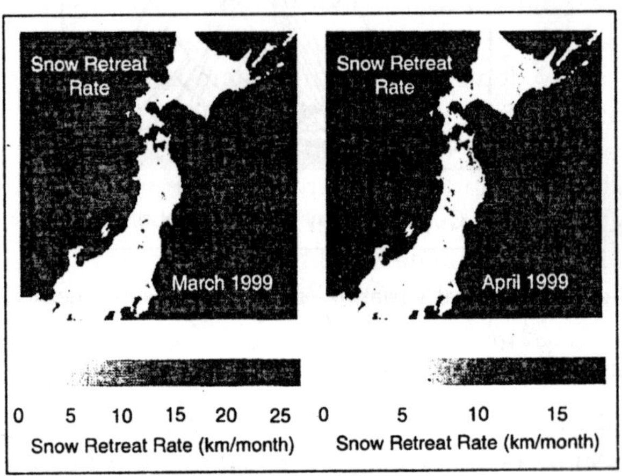

Fig. 2.4 Snow Retreat Rate in Snowmelt Season

The Relationship Between Snow Retreat Rate and Snow Increasing Coefficient

The distribution of snow increasing coefficient and the distribution of snow retreat rate is estimated in previous sections. The movement of snow line in snowmelt season is evaluated with these databases. Figure 2.5 shows the relationship between snow retreat rate and snow increasing coefficient at previous month. Many plots concentrate in the neighborhood of origin of coordinates. These snow depths are about zero, therefore snow retreat rate and snow increasing coefficient is small. Omitting these plots, snow retreat rate is fast in region the gradient of snow depth to elevation is low and contrary snow retreat rate is slow in region it is high, in other words, the relationship between snow retreat and snow increasing coefficient is inverse proportion.

The reason is that if the gradient of snow depth to elevation is high, the snow volume in the range which snow line retreat is fast and snow retreat rate is slow. On the contrary, snow volume is low and snow retreat rate is fast, if it is low. In Figure 2.5, symbol A indicates the value at Iwamizawa, (43.395°N, elevation 6m) and B does at Ishikari, (43.21°N, elevation 31 m).

These points are at Ishikari plain in Hokkaido Island. Snow retreat rate is fast and snow increasing coefficient is low because snow depth is low but climate is warm. C does at Aterazawa, (38.36°N, elevation 140 m) and D does at Inawashiro, (37.56°N, elevation 527 m). These points are in Yamagata valley and Inawashiro valley. Snow retreat rate is slow and snow increasing coefficient is high because snow depth is high but climate is cold.

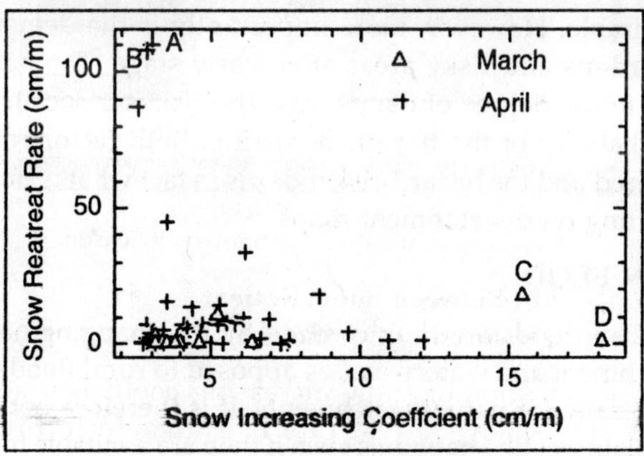

Fig. 2.5 Relationship between Snow Increasing Coefficient and Snow Retreat Rate

SUMMARY

The summary of this report is followings. At first, the altitude distribution of snow depth at about 200 AMeDAS snow depth point is evaluated. As results, the gradient of snow depth to elevation (snow increasing coefficient) is lower in Japan Sea side than in Pacific Ocean side. In addition, it is higher on low elevation area than in high elevation area. Secondly, snow retreat rate is estimated and relationship with snow increasing coefficient is evaluated. After all, it is possibly that snow increasing coefficient is inversely proportional to snow retreat rate in snowmelt season.

HAZARD-RISK INDEX FOR URBAN FLOODING

Urbanization is becoming a fact of life and problem for many population groups, particularly in developing countries. The level of migration to urban and economic centres in many cases outstrips the rate of earning and servicing. The result is that there is great pressure for land, in particular for low cost housing and in many cases, informal housing clusters developed faster than planning procedures can advance. There are often catastrophes when floods occur or due to heavy pollution of the urban waterways. Therefore low cost management systems are desirable. However, more importantly is the demarcation of hazardous and risky areas at an early stage. The hazard is related to the degree of danger and the risk is associated with the probability of the hazard occurring. Both factors must be considered and the hazard-risk index is in fact what is proposed for plotting on development maps.

URBAN FLOODS

Urban floods are characterized by the changing nature of the catchment and waterways, as opposed to rural floods which are generally for natural catchments. It is therefore not easy to extrapolate existing gaugings even if they are available for urban streams because the characteristics, in particular the intensity of the flood and the management of the flows may change over time. Indeed, flood management by way of detention or retention is more important for urban streams as the floods increase due to urbanization. Not only is there an increase in impermeable cover in the case of urban development, but also the impermeable cover such as roofs and roads is in many cases directly connected to the drainage network. The drainage network increases flow velocities due to lower friction factors and deeper channels. The consequence is that the critical storm duration is shortened compared with natural catchment and therefore a more intense design storm becomes the norm. This means that the spectrum of floods changes due to urbanization.

It is however often easier to install management structures in urban streams as the streams are smaller, but more particularly the economics dictates that management systems should be installed. Therefore it is necessary to conduct rainfall-run-off modelling when studying urban water systems in order to optimize the level of management and in order to calculate the water levels associated with different frequencies or occurrence or exceedance. The analysis of rainfall records often yields long-term records statistics and more particularly rainfall intensities for different duration storms which can be built into a model for rainfall-run-off calculations such as Xp Swmm Or Rafler..

STORMWATER MANGEMENT AND RISK

The design process from hereon is iterative as the level of management such as by detention or diversion storage influences the extent of flooding downstream as well as the backwater of the reservoir. In socio-economics optimum level of management for a selected time horizon is necessary. It should also be borne in mind that entire analysis should be done for future level of urbanization when flood intensities and volumes are higher.

The next stage in the method of analysis is to calculate flood levels using a hydraulic–type model such as RIVERCAD or BACWAT. This requires contour information which can be obtained by digitizing contour maps or aerial photography. The resulting computed water levels can be replotted on digitized maps on a GIS system.

The resulting maps and cross-sections through the channels will indicate different water levels with different recurrence or risk of exceedance. That is, the lowest waterway is plotted usually that associated with a 1–year flood whereas the channel thalweg is that associated with the average minimum dry season flow. Successively higher levels will cover the flood way and the flood fringe.

The waterway is usually taken to be that within the bounds of the 1-year flood, whereas the flood way is taken to be that within the 20–year flood line and development is seldom prohibited below this level.

The next higher level such as the 100–year flood could embrace the flood fringe and again only certain development would be permitted within this fringe subject to the sensible design and an authoritative consideration of the effects of construction within the waterway or flood fringe. *I.e.* the more development that occurs with these fringes, the greater the backup effect is, and therefore the wider the waterway and flood fringe become.

It is often not easy to anticipate the type of development which will occur within these lines. The different probability or risk zones are demarcated with a risk index. That is, above the 100-year flood line will be given a zero rating, between the 20–year and 100–year flood line will have a rating of 1, and below the 20-year flood line, the risk index will have a value of 2.

HAZARD INDEX

The level of flooding which can be tolerated depends on the socio-economic impact of the flooding. Thus there may be no hazard in rural areas if there is no people danger within the flooded area. The danger generally increases with deeper flows and faster flow velocities. Thus, the ideas developed in Minnesota were formalized by the New South Wales authority who indicated values of velocity x depth as giving an index.

Thus a velocity of 1m/s and a depth of 1m would result in a high hazard as buildings could be washed away. A value of the product of 0.5 could still be dangerous to people, particularly children. An even lower water depth, that is of the order of 200mm, could cause severe damage within buildings, but may not be a danger to life. This level may also be manageable provided a warning system exists and people can use sandbags or other diversion means to avoid extensive economic damage. A water level as low as 0.05mm would be a danger to traffic on roads.

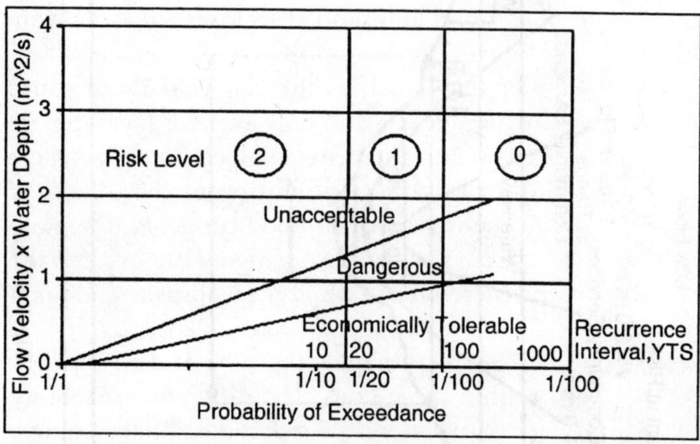

Fig. 2.6 Flood Risk Diagram

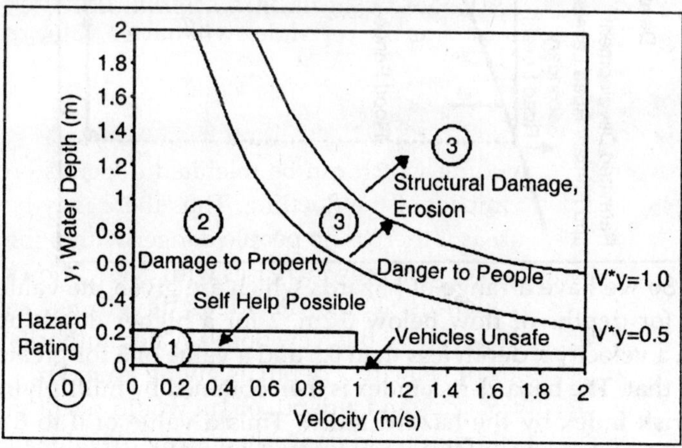

Fig. 2.7 Flood Hazard Diagram

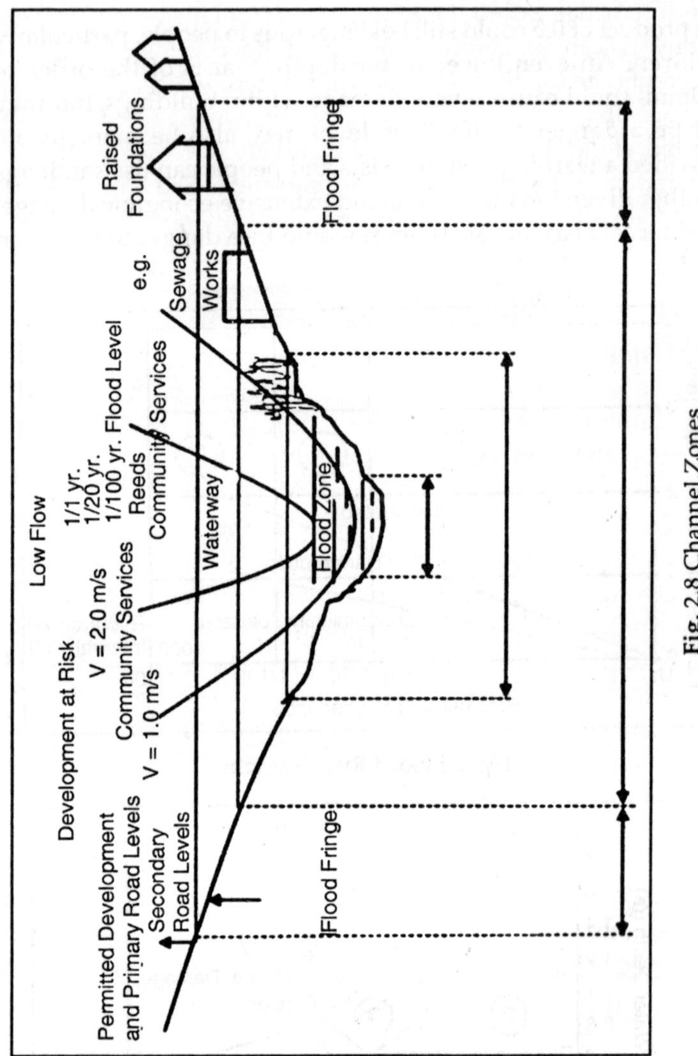

Fig. 2.8 Channel Zones

So we have a range of hazards which are given the values of 1 for depths of flow below 0.2m, 2 for a higher depth but with a velocity x depth less than 0.5 and a value of 3 for greater than that. The hazard-risk index is then obtained by multiplying the risk index by the hazard index. Thus a value of 0 to 6 is possible, where 0 is of little concern and 6 is of highest concern.

The authorities would therefore obtain from the GIS system the hazard-risk index before deciding whether to allocate land to different types of development. Thus hazard–risk indexes above 3 generally require constructed waterways and/or barriers erected along the fringe. For successively decreasing hazard–risk index, different levels of development and in particular raised development may be tolerated but not encouraged. Careful monitoring of the fringes is therefore required.

RISK ANALYSIS AT DAMS

Failure of a dam occurs when one or more of the failure events occurs, such as overtopping, seepage, piping, instability, etc. Numerous studies of dam failures have indicated that overtopping is a major failure mode of earth and rockfill dams. The average dam failure probabilities for (all kind of) dams was found to be approximately 10^{-3} per year per dam. In thirty per cent of all cases overtopping is the cause of earth dam failure.

Overtopping occurs when the water level of the reservoir behind the dam rises above the dam crest. It may result from failure to make timely and adequate releases of floods through the spillway and flood release outlets, wave action induced by wind, landslides, earthquakes, and other geophysical forces, and the combination of these effects. In the context of this paper only flood and wind events are considered for the evaluation of the overtopping probability of the dam.

Hydrological phenomena like precipitation or run-off always appear as multivariate events. For many problems in water management it is not satisfying to know only the frequency of single characteristics. It will be more important to know the probability of the whole event. Therefore they should be characterized by numerous parameters.

Dam failure is not a determinable event and hence quantitative evaluation of dam safety requires probability theory. The proposed probabilistic design concept for spillways of dams includes a multivariate statistical model for the simulation of incoming flood events, which is able to describe the hydrological process entirely.

CURRENT DESIGN PRACTICE

The actual hydraulic design practice in Austria is to select a definite load case (or several cases, if necessary) for which the calculation is made. This design concept deals with only one random variable, named the "peak discharge" of the flood. The design flood itself is defined as the peak run-off (anunivariate value), which can be discharged by the spillway outlets under special (mostly fixed) conditions, without endangering the dam or important parts of the construction.

Other parameters of the design event are more or less, as suitable assumed. Currently operating states due to reservoir management, do not find any consideration in the official water law procedure. (the initial water level in the reservoir at the beginning of a flood event or the condition of the crest and spillway gates and the bottom outlet or the wind depending freeboard height) and are specified as fixed values.

The uncertainty in the estimation of design floods induced the water authorities to set up safety principles for the dimensioning of flood spillways, which seem to be relatively highly compared to other endangering potentials. Mostly it has not been considered, that there exists a spectrum of hydrographs for each probability, from which the decisive event and the resulting peak flow has to be found out, while considering the retention.

PROPOSED SOLUTION

These are reasons to develop a probabilistic design concept for spillways, which is taking account of the demands of safety (against uncontrolled overtopping of dams) as well as of economy due to reservoir management. Nevertheless it has to be considered, that the safety of hydraulic structures does not only depend on the chosen computing method and the choice of the design probability, but also on the quality of the available hydrological data.

Especially the choice of the design probability varies in the countries of the world. The return period of the design events for large storage reservoirs vary from a 100 years up to 10.000 years and is limited in most countries by the probable maximum flood.

MAIN LOAD VALUES

The proposed method, originally developed by Pohl, R. at the Dresden University of Technology, is standing out by the fact, that the computed water level in the reservoir is consequently the only design value for evaluating the overtopping safety of the dam.

This design value is resulting of three main load values, namely the:

1. Random initial reservoir water level at the beginning of a flood event...h_0
2. Additional elevation height due to a random flood event (inflow)...h_r
3. Required freeboard height due to a random wind event...h_f.

According to this, the design value (=water level) itself is also a random variable, which can be estimated by means of the method of statistical trials. An exact solution is difficult or impossible, due to flood routing and freeboard calculations. Within this design concept, the "failure" is defined as a non–desired hydraulic state, independently of damages to property or personal injuries.

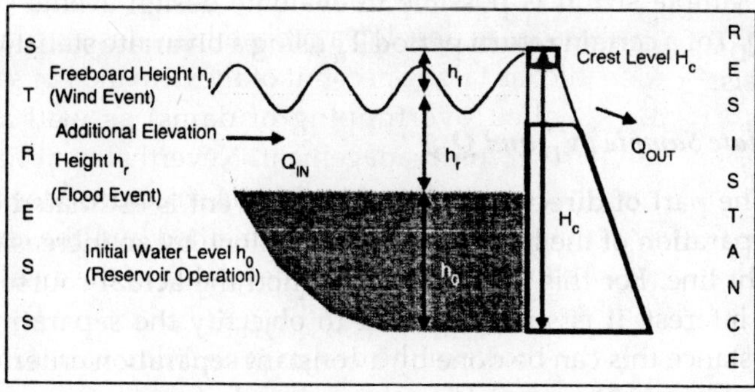

Fig. 2.9 Definition Sketch, Approach of the Overtopping Probability

Initial Reservoir Water Level (Reservoir Management)

At the beginning of a flood event an initial reservoir water level h_0 is to be found, which is mathematically considered to be random.

The separate entry of the initial reservoir water level guarantees the economy of the design concept and is representative for previous floods or droughts, the reservoir operation and the flood management, the availability of crest and spillway gates and the availability of the bottom outlet and the penstocks.

Additional Elevation Height (Flood Event)

Of large influence for the safety against overtopping of the dam is the incoming flood event. The retention of the flood hydrograph in the uncontrolled storage normally rises the level in the reservoir.

For this reason the flood event itself has a special meaning within this design concept. A bivariate flood model has been developed at the Institute for Hydraulics and Hydrology at the Graz University of Technology and should become a part of the design concept.

Characteristics of Flood Events

The peak Q_D and volume V_D of direct run-off are the best describing characteristics of a flood event. With a sufficiently large sample size it is possible to evaluate design events (V_D and Q_D) of a certain return period T_n using a bivariate statistical analysis.

Bivariate Sample (V_D and Q_D)

The part of direct run-off of a flood event is estimated by the separation of the base flow, which is done by an increasing straight line. For this quantitative method the actual course is of no interest, it is still important to objectify the separation. For instance this can be done by a constant separation criterion ($V_D/V_{B,1}$ = const.).

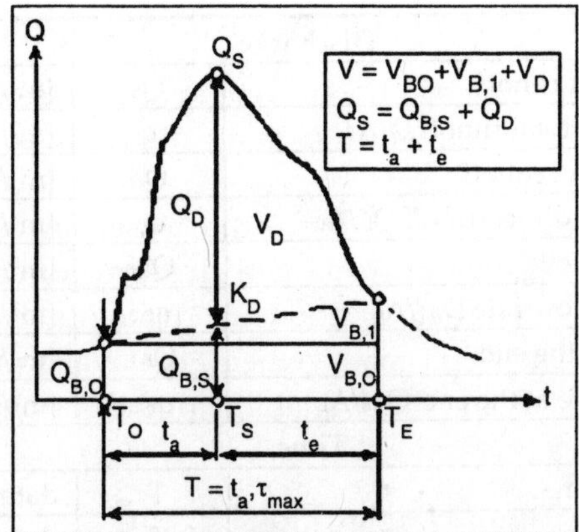

Fig. 2.10 Schematic Representation of a Flood Event

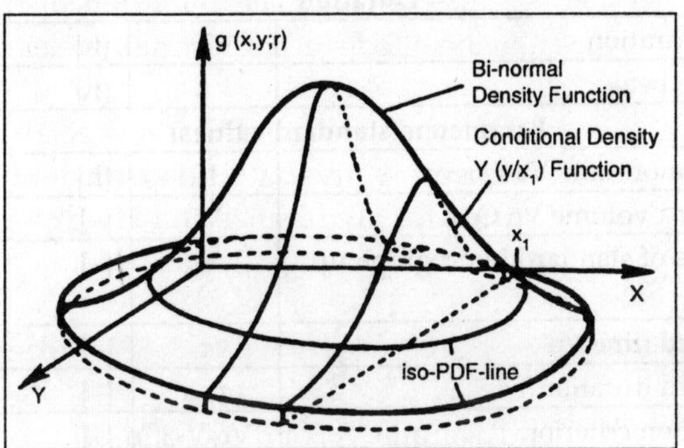

Fig. 2.11 Bi-Normal Density Function

Table. Flood Characteristics

Area		
Basin area	A_E	[km²]
Volume		
Total flood volume	V	[m³]
Height of total runoff V/A_E	H	[mm]
Volume of direct runoff	V_D	[m³]
Height of direct runoff V_d/A_E	H_D	[mm]
Initial baseflow volume	$V_{B,0}$	[m³]
Separated baseflow volume	$V_{B,1}$	[m³]

Discharge		
Peak of total runoff	Q_S	[m³/s]
Peak rate of total runoff Q/A_E	q	[m³/s. km²]
Peak of direct runoff	Q_D	[m³/s]
Peak rate of direct runoff Q_D/A_E	q_D	[m³/s. km²]
Initial baseflow	$Q_{B,0}$	[m³/s]
Initial baseflow rate $D_{B,0}/A_E$	$q_{B,0}$	[m³/s. km²]
Baseflow at the end	$Q_{B,E}$	[m³/s]
Baseflow rate at the end $Q_{B,E}/A_E$	$q_{B,E}$	[m³/s. km²]
Time		
Time to begin	T_0	date, time
Time to peak	T_S	date, time
Time to end	T_E	date, time
Duration		
Total duration	T	[h]
Time to peak	t_a	[h]
Parameters(standard values)		
Peak runoff time V_D/Q_D	t_m	[h]
Standard volume $V_D/Q_D.t_a$	VS	[-]
Volume of standard hydrograph up to peak	Vs	[-]
Standard time t/t_a	τ	[-]
Standard duration T/t_a	τ_{max}	[-]
Separation criterion	$V_D/V_{B,1}$	[-]

Bivariate Distribution Function

To prevent an escalation of variety in methods, the bivariate normal distribution function (Fig. 2.11) is used exclusively. An excellent fit can be achieved by a transformation of single samples into normal distributed samples. The easiest way to get an approximate "normalization" is to transform the single samples in a way that the coefficients of skewness become zero.

Lines of Equal Probability Density Function

First of all it is possible to compute the ISO-PDF-lines, which are limiting certain random areas. They are horizontal cuts through the bi-normal PDF, which are ellipsis in the normalized V_D, Q_D–system. With the help of these lines is possible to make a semi-graphical test of fit.

"Design curves" - Lines of Equal Return Period

The probability of an event occurring in a certain V_D, Q_D - area is equivalent to the volume above this area up to the bivariate PDF. There are theoretically infinite possibilities to delimit such an area for integration. The only efficient definition of a bivariate flood probability is the so called "upper right probability" pur which is the probability of the occurrence of a pair of values V_D, Q_D in the upper right quadrant. An ISO-$_{pur}$-line is a horizontal cut through the bivariate distribution function relating to the "upper right quadrant". It is also possible to define an ISO–$_{pur}$–line as a line of equal return period or as a "flood design-curve". For different return periods a spectrum of design-curves is obtained. The intersection points of an ISO-$_{pur}$-line with the V_D- and Q_D-axes corresponding to a certain return period Tn are the respective values V_n and Q_n. The course within these points respectively within the event-range has nearly the shape of a quarter-ellipse (regarding only bivariate flood analysis). So design-curves can be approximated as quarter ellipses after the collection of flood-peaks for certain return periods Q_D [m³/s], volumes of direct run-off for certain return periods V_D [m³], a typical local standard hydrograph or standard hydrograph-range, typical local range of the time index tm = V_D/Q_D [hours], and mean base flow or initial base flow $Q_{B,0}$ [m³/s] corresponding with the increase of base flow $V_D/V_{B,1}$ [–].

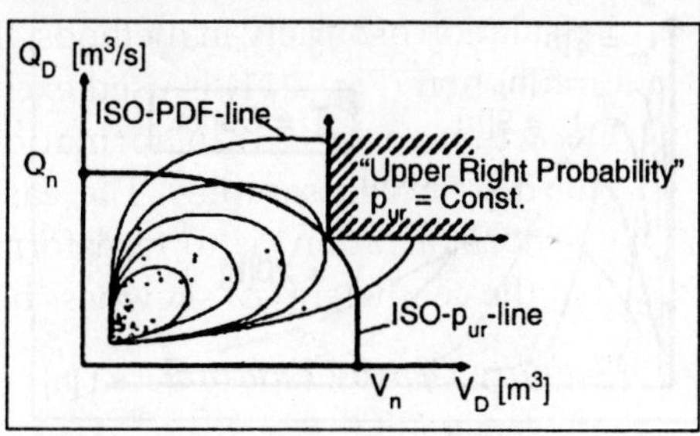

Fig. 2.12 ISO-PDF-LINES

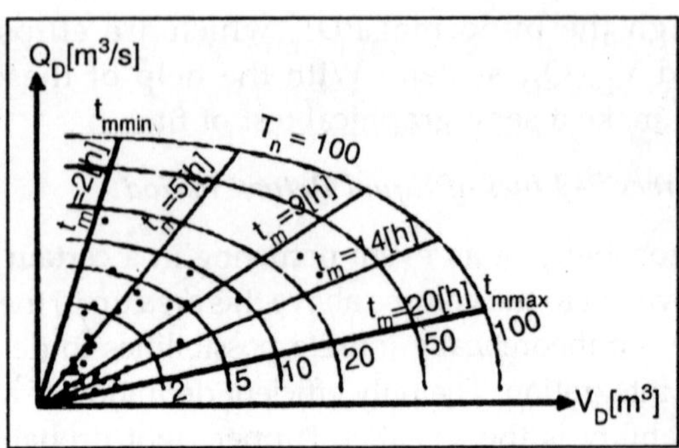

Fig. 2.13 Curves of Flood Events of Equal Return Period

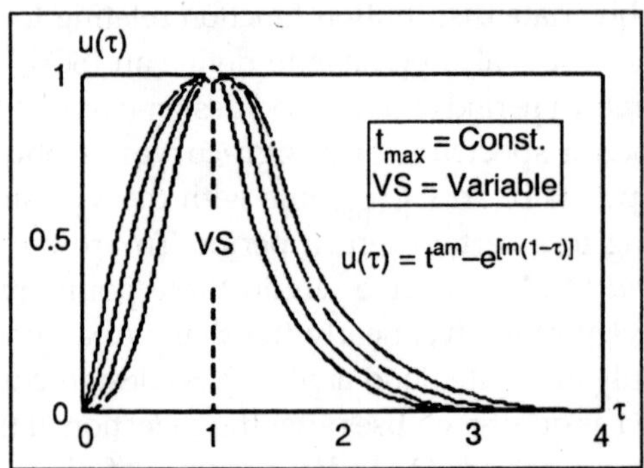

Fig. 2.14 Standard Hydrograph

Standard hydrograph - event range:

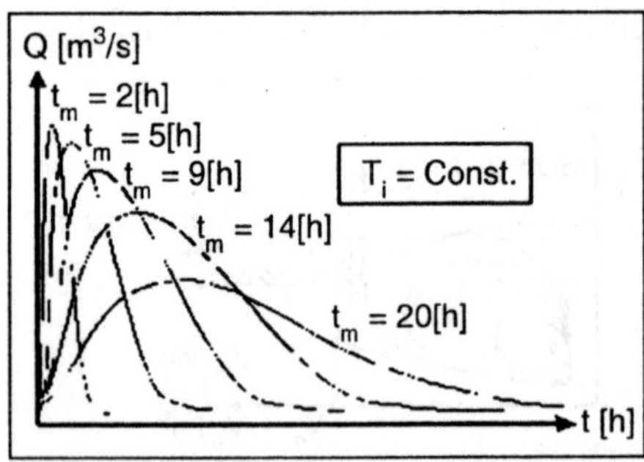

Fig. 2.15 Spectrum of Design Flood Hydrographs

A direct run-off hydrograph standardized to the peak u= 1 and the time to peak τ= 1 is called standard hydrograph. It is necessary for the model, to reduce the multivariate "flood" process to two dimensions. The chosen analytical 2-parametric function makes it possible to choose the type of standard hydrograph in a wide range according to the basins properties. To obtain the standard parameters (a, m), direct run-off is separated from measured flood events and thenstandardized, by the determination of the standard volumes vs and the standard times τ_{max}= T/t_a.

There are no floods with an extreme peak run-off and a volume equal to zero. For this reason, there always must be a certain "event range", which is possible to delimit approximately with straight lines through the origin. This straight lines are defined by the time index t_m.

The time to peak t_a of an event point lying on a design curve results directly from the value of the time index t_m and the assumed standard hydrograph. Therefore the boundaries tm,min and $t_{m',max}$ can be estimated by the probable minimal and maximal values for t_a for a certain watershed. The course of the "design-curves" is similar to a quarter ellipse only within the range of tm,min and $t_{m,max}$.

Design Flood Hydrographs

After construction of the design curves for different return periods, the events along a "T_n-years" line are converted into a "spectrum of design flood hydrographs" with the constant return period T_n.

Freeboard Height

The calculation of the freeboard height as shown in the 3rd column of Fig 2.15. widely corresponds to the guideline 246/ 1997 of the German Association of Water Resources Management and Land Improvement. Both wind set–up in the reservoir and wave run–up on the slope of the dam are considered. They are functions of the wind direction, fetch, water depth, and wind velocity.

FLOW CHART OF THE CALCULATION PROCEDURE

The flow chart of the calculation procedure is schematically shown in Fig 2.14. In this procedure the flood event with the maximum discharge peak of n arbitrary years is simulated n–times using random numbers, whereby the overtopping probability is distinguished, concerning the direct probability of overtopping p_2 (only due to flood meeting the initial water level) and the indirect probability of overtopping p_3 (due to flood and wind effect). A practicable design software has been developed to estimate hydraulic–hydrologic failure situations by means of the method of statistical testing and includes the multivariate flood model described before.

CASE STUDY AND RESULTS

Salza dam impounding the river Salza is an annual storage in Austria has been chosen to explain the results of the proposed design method. The 53 [m] high constant-angle arch dam was built between 1947 and 1949 and covers a catchment area of A_E = 150 [km²]. Salza dam was the first arch dam of medium size in Austria. This is the reason why an excellent flood data base is available for any kind of investigations. The overtopping probability of the dam crest in an arbitrary year is between 0,7 and 15 [%] due to flood event, respectively flood and (the assumed) wind event together (85 and 87 [%] in the design case).

A specific feature of the Salza dam is its ungated spillway crest over the whole crest length (121 [m]) which, rising towards the flanks from elevation 771,5 to 772,5 directs overflowing water into the middle portion irrespective of the flow rate. So the Salza dam is rather insensitively against overtopping and this is the reason why such high overtopping probabilities are acceptable. Actually, priming of the spillway occurs several times a year at the Salza dam.

Calculation assumptions to the case study:

- The initial water level was uniformly distributed between normal top water level and 8 [m] below normal top water level
- The bottom outlet is closed during a flood event

- Because of missing wind data at the Salza dam site,wind velocity was assumed with 15 [m/s]

Table. Results of the Overtopping Probability of the Salza Dam in Austria

Overtopping Probability [–]					
		Arbitrary Year		Design Case (Q ³ HQ5000)	
Critical water level [–]	Altitude [m a.s.l.]	p2 only Flood	p3 Flood + Wind	p2 only Flood	p3 Flood + Wind
Normal top water level	771,0	1,0	1,0	1,0	1,0
Overflow crest	771,5	1,0	1,0	1,0	1,0
Dam crest	772,5	0,007	0,15	0,85	0,87

Further results of another case study concerning a combined concrete gravity and rockfill dam with an controlled spillway using the proposed method are reported by Hable, O. *et al.*. By applying the proposed method, it is possible to estimate the correlation between the overtopping probability and a freely selectable "critical water level" (the maximum storage level in the reservoir, the crest level or the overflow crest of the spillway).

Therefore prognoses about the different probabilities (of keeping of the storage level, of overtopping of the dam or of priming of the spillway) are possible. Beyond this it is possible to dimension spillways of hydraulic structures (not only dams) by matching the demands of safety as well as of economy. For this reason the main advantage of this probabilistic design method including the proposed multivariate flood model are an almost complete and realistic processing of the available information, an excellent simulation of the incoming flood event, a risk estimation tool as a base for taking measures if necessary and the comparability of dams with different spillways, storages, types of constructions and purposes concerning their security. A research project together with the association of the power plants of Austria has the aim to develop and improve the proposed method. It should be possible to open up new paths in the field of design works, if a reorientation in finding of design values is made possible in the official water law procedure.

WATER RESOURCES MANAGEMENT SYSTEM ORIENTED

In China, river basin management has existed for hundreds of years. Currently, river basin management organizations have been created for important rivers and a system has taken shape that water resources of large rivers are managed within a basin framework. Changjiang (Yangtze) is the longest river in China. Execution of river basin management in a modern concept dated back to the 1920s. By 70 years' efforts, especially through vigorous development since 1949, great achievements have been accomplished in every undertaking in the valley, and the strategies for river basin management have been improved substantially. Nevertheless, with social and economic development in the valley, the current river basin management system can no longer sufficiently fit the new situations. With a view to accommodate to the needs in the new millennium, it is necessary and urgent to establish a new management system on basis of the current one. This new system shall be a guarantee for preventing flood and drought disasters in a more efficient manner and promoting the integrated and sustainable development of water resources in the Changjiang valley.

CURRENT SITUATION OF CHANGJIANG RIVER BASIN MANAGEMENT

Currently, water management for Changjiang River valley applies such a structure, which is a combination of basin management by the central government and regional management by local governments. Changjiang Water Resources Commission (CWRC) is the river basin management organization, dispatched by the central government to execute water administration function in the Changjiang valley. In accordance with the principle of unified management combined with departmental management, the CWRC has the responsibility to manage water resources and all river channels within the scope of the valley on basis of related laws and regulations. Its missions include: comprehensive harnessing of water resources, administration and management of backbone water projects, long-term river basin planning, water administration, water development coordination, supervision and service-providing and the integrated development, utilization and protection of water resources.

Local governments within the valley can carry out regional water management schemes tailored to the actual conditions under its administrative jurisdictions according to related regulations. It is an important component of river basin management and the materialization of unified management combined with regional management. At present, the main tasks assumed by the CWRC in harnessing, development and management of water resources in the Changjiang valley and related problems are as follows.

The CWRC has been Working For the Establishment of a Sound Water

Water administration department is the core department for river basin management in the valley. The main tasks in this field are the following: publicizing related laws, regulations and policies and supervising their execution; participating in the formulation of and amendment to related laws, regulations and policies; resolving inter-provincial water affair disputes; carrying out water resources management and river channel management; executing the law for water administration and other related activities. The primary problems involved in this field are as follows: Institutional and policy weaknesses concerning the legal position and administration functions of the river basin management authority, which needs to be strengthened; Shortage of efficient means for this authority to coordinate benefits between different regions and different sectors.

Flood Prevention and Disaster Reduction

Flood disaster is the most severe natural hazard in Changjiang valley, always with huge flooding losses. Flood prevention and disaster reduction have been the priority task for the CWRC. Through continual construction of flood control projects, flood control standards on Changjiang River have been raised significantly.

Since the year of 1954, in which an extraordinary heavy basin–wide flood occurred, the main dykes have withstood all the heavy floods happened on the River ever since, the 1998 floods in particular, which has demonstrated the great achievements in flood control engineering constructions. Safety for hundreds of millions of people' properties and lives has been ensured and the continual and stable growth of national economy guaranteed. Nevertheless, due to the fact that ground elevations in the plain areas on the middle and lower Changjiang are generally lower than flood stages of the River and that flood control standards are in general on the lower side, flood control situations on the Changjiang have always been severe.

Water and Soil Conservation has been Implemented With Force

Attention has also been paid to both ecological system protection and water and soil conservation. Up to now, more than 100,000 sq. km areas of water and soil loss regions have been brought under control. In these conservation areas, water and soil loss has basically halted and environment quality improved. However, water and soil loss in the valley has not been completely brought to an end. It is estimated that the amount of area existing water and soil loss is about 560,000 sq. km in the valley. The areas are still increasing due to human activities such as deforestation. Sediment load discharged into the sea reaches 500 million tons each year.

Great Importance has been Attached to the Fundamental

Hydrologic and survey data collection and processing have

provided basic data for planning, scientific researches, design, flood fighting and water administration and basis for decision making. Currently, there are more than 6400 hydrologic and meteorological stations in the valley with many survey and experimental units scattered all over the valley. However, great difficulties have always been encountered for these units to carry out their works due to institutional and market factors.

These difficulties include fund shortage, too heavy burdens of various kinds, difficulty in translating scientific and technological advances into production forces and competition without order. All this has severely constrained the smooth implementation of fundamental programmes in the valley.

River Basin Planning has been an Important Task of the CWRC

During the mid-1950s, the CWRC carried out overall river basin planning in the light of requirements for economic development at the time. In 1959 the CWRC formulated the Report on Main Points of Changjiang Valley Comprehensive Planning, which has been the guidance for the development, harnessing and management of the River. In the years that follow, the CWRC has completed designs for Three Gorges Project, Danjiangkou Project and so on, planning for river reaches of the main stem and the primary tributaries and planning for South to North Water Transfer, flood control on the middle and lower Changjiang River and river training and so forth.

This report has been approved by the State Council in December 1990 and made the basis for the development, utilization and protection of water resources and flood prevention in the valley. However, there is a shortage of definite legislation for supervising the implementation of river planning so as to ensure the execution of the planning. It happened now and then events that violated the planning principles. There exist weaknesses in functions for the river basin management authority to execute macroscopical management and supervision.

The Execution of Water Management for Protection

Up to now, the CWRC has received 75 pieces of water withdraw register applications and granted 45 pieces of water withdraw licenses, the annual approved withdraw amount totaling 374.1 billion m^3. Through implementation of water quality monitoring and protection measures, the discharging of pollutants in the valley is being brought under control.

Great efforts have been made to realise unified management of water quality and water quantity so as to guarantee the sustainable development of water resources in the valley. However, due to the fact that water management operations have not been completely put into place and that water quality management system has not been tidied up yet, water management in the valley has been severely hampered.

CWRC has Further Provided Services to and Participated in Social and Economic

Upon the firm footing of carrying out survey and exploration, design and scientific researches for water resources and hydropower development, the CWRC has actively involved it in other fields, such as construction, transportation, information, commerce, trade and so on. It has made unique contributions to water resources and hydropower development, especially the construction of large–scale backbone projects in the valley, contributing a great deal to social and economic growth. However, the current market of survey and exploration, design and scientific researches is not adequately mature yet, and related regulations have not been made sound and complete, which has influenced the execution of these activities. In addition, for lack of fund and policy support from the government, the river basin management authority is hard to organize construction of backbone projects on the main stem and its primary tributaries and to carry out rolling development planning. It is prescribed in a State Council approved responsibility-division scheme that river basin management authorities shall assume the responsibility to organize the construction and to manage the backbone water projects or key inter–provincial projects. Actually this provision is just idle theorizing, because the river basin management authorities have no means to develop their mainstay businesses, say no more of rolling developments.

EXPERIENCES FROM OVERSEAS RIVER BASIN MANAGEMENT PRACTICES

The concept of unified water management on a basin basis has been accepted widely. Currently, there are success cases in implementing river basin management in the world. Included are the Tennessee Valley Authority (TVA) pattern in the U.S., the Rhone River Corporation management pattern in France and the Thames Water Affairs Authority pattern in U.K. All these management patterns have one point in common. That is, they have established sound management systems tailored to its actual conditions and achieved success in executing river basin management.

The TVA Pattern

The TVA was created by the U.S. Congress with the advice from the president for the purpose of developing and harnessing the Tennessee valley. It is an independent federal corporation with strong administration power and the flexibility and creativity of a private company type. The TVA's aim, objectives, operation scope, rights and responsibilities are definitely specified by legislation.

The TVA has multiple objectives. With a view to reduce contradictions caused by the duplication of functions with other executive agencies, the TVA was granted relatively high independence of self–rule. The TVA is administrated by a board of three officials who are appointed by the president, subject to the approval by the U.S. Senate. All the rights are in the hand of the Board, including personnel management, finance and price setting. The officials selected to the Board were generally experts of related fields. All the important decisions are collectively made by the Board. In order to facilitate growth of the TVA, the federal government granted to the TVA many privileges. A few large–scale enterprises and power stations were transferred to the hand of the TVA. Its properties and incomes were exempted from taxation. Government guarantee was granted to the TVA to facilitate the issuing of business bonds to support construction of key projects and so on. Under unified management of the TVA, striking success in flood control, transportation, power-generation, agriculture, forestry, fishery and tourism has been achieved through more than 60 years comprehensive development in the valley.

The success has been recognized world widely and the TVA was imitated in many countries with the creation of similar river basin management agencies.

They have many in common:

- The establishment of such agencies is aimed at improving economic conditions of river basins. And the agencies are granted a wide range of powers for promoting social and economic development in the basins.
- These agencies are independent executive departments of the governments and are responsible directly to the central governments.
- The agencies are of high self-rule independence, which is specified through legislation.
- Special funds are appropriated for start-up development and follow–up development is supported. Corresponding policy supports are provided.

The Rhone River Corporation Management Pattern

With a view to coordinate relations and benefits of different districts and different sectors during development and harnessing of the Rhone River, the legislative body of France enacted a law in 1921 creating the Rhone River Corporation (RRC). In the mean time, chartered concession was granted to the RRC to develop and manage the Rhone River.

The RRC is a joint-stock public utility company controlled by French government. Its main missions include hydropower generation, navigation and irrigation in the basin and development of riparian land. When executing its functions, the RRC is the owner, the contractor and the manager. At present, the RRC has 320 stockholders, including local governments related directly with development of the basin, public and semi-public organizations, such as the State Power Corporation and the National Railway Corporation. A board of directors was set up in the RRC, chairpersons of it are cabinet members, generally politicians. The board is consisted of 30 persons, seven from related executive departments of the government, five stockholder representatives, five staff representatives, five experts and 8 beneficiary representatives. The board calls a meeting in every month, discussing and making decisions on important issues and the general-manager's report to the board. Under the board is the general-manager, who is in charge of daily affairs of the company.

During the time period of more than half a century, cascade development of the 500–km river channels in the French territory has been completed, as planned in river basin planning. The Rhone River has been one of the most fully developed rivers in the world. The basic experience is as follows.

The comprehensive development is based on reasonable planning, favorable policies and modern management rules. Hydropower exploitation is the mainstay business, which is made the base for developing and harnessing the whole river and realizing follow-up development. Full considerations have been given to the comprehensive development and conservation of water and soil resources in development planning. Flood control, hydropower generation, navigation, water supply and environment protection are equally considered.

Policy support is available from the Government. The RRC was granted the chartered concession to exploit and manage the Rhone River and the riparian land for 100 years. As a nonprofit public utility agency, the RRC's incomes of various kinds are exempted from taxation. The company does not need to distribute dividends to large amount of local shares. Resources fees of several types are collected by the RRC to make up fund deficit. The company has the right to determine sale prices for its power supply by considering the operation and maintenance cost, payment of loan and interests and ensuring reasonable profits so as to guarantee good performance of the company. Scientific management rules and modern operation tools are made use of. All this has made it possible for a limited number of employees to manage a huge system in an efficient manner, including water administration, project construction, operation and maintenance and scientific researches and so on.

Thames Water Affair Authority (TWAA) Management Pattern

The TWAA is one of the 10 water affair authorities established in the U.K. according to the Water Act of 1973. Its main missions include formulation of long-range water management strategy and plans; development and construction of new water sources; providing with clean and adequate water to residents under its jurisdiction; sewage processing; restoring and safeguarding the cleanness and sanitation of the river and other water bodies; maintaining, improving and expanding freshwater fishery; full supervision of land drainage and flood control under its jurisdiction; providing protection for recreation water bodies and related land; investigation of integrated water management situations under its jurisdiction and working out medium- and long-range water supply plans.

The executive body of the TWAA is the board of directors. The chairperson of it is appointed by the Environment Secretary (ES). Members compose of two parts. One is the persons appointed by the ES, and the Agriculture, Fishery and Food Secretaries. The other part is local representatives. The number of representatives appointed by the government shall not outnumber that of the local ones. The board is responsible for working out long-range strategies and policies and employing the general–manager for the daily works.

The water affair authority (WAA) is a non-government agency. Actually it is a highly independent public utility organization that assumes sole responsibility for its profits or losses (flood control and drainage are funded by appropriations from governments of all levels). The WAA has the right to determine charging standards for its services and collecting fees. The basic principle is to ensure a balance between its incomes and expenditures plus prescribed profit rates. Fees collected include water fees, sewage fees, drainage fees, environment protection fees and water resources fees and so on.

ESTABLISHMENT OF A NEW WATER RESOURCES MANAGEMENT SYSTEM

To sum up, the establishment of a new water management system oriented to the 21st century is of extremely importance to realise sustainable development of water resources and promoting speedy and stable growth of society and economy in the valley.

This new management system shall have such basic characteristics as follow:

- The river basin management agency shall be more powerful.
- Management contents shall be extended to include all the fields of water resources, which shall be wider than the existing items.
- The responsibilities and rights of river basin management and regional management shall be more definitely divided.
- Regional management shall be subordinated to river basin management and sector management subject to integrated management.

The river basin management authority shall be the main part assuming the responsibility of river basin management and development. Backbone water-resources development programmes in the valley shall be controlled from the very beginning to the end by this authority on behalf of the government, so as to make the comprehensive strength of this authority continually grow.

Therefore, the newly established river basin management structure for Changjiang valley shall have the following function systems:

- A more perfect water administration system;

- Solid fundamental programme system;
- High-level scientific research and consultation-providing system;
- Multi-level finance support system;
- Perfect project approval, construction and management and quality monitoring system; and
- Flood disaster prediction, forecasting, monitoring, assessment and emergency handling system.

Establishment of a Powerful River Basin Management Authority

To strengthen river basin management, a powerful agency need to be created, which can be called, for example, the Changjiang Valley Water Commission (CVWC) or Changjiang Valley Sustainable Development Commission. This commission can be headed by the leader of the State Council and compose of chiefs from related ministries and departments, the provinces, autonomous regions and municipal cities in the valley. This commission will assume full responsibility for development and utilization of water resources, flood prevention and disaster reduction and environment protection in the valley.

The establishment of such an agency shall be an important organizational guarantee for integrated water resources management in the valley. At present, the Changjiang Flood Control General Headquarters and the Upstream Changjiang Water and Soil Conservation Commission have already been established.

The head of the Committee can be concurrently held by the Premier or a vice-premier. Committee members can compose of chiefs of the State Development and Planning Commission, Ministry of Water Resource (MWR) and related ministries and departments and the chiefs of provinces (autonomous regions and municipal cities) in the valley and the Commissioner of the CWRC, who can also concurrently hold the post of Secretary–General.

A so–called Office of Changjiang Valley Water Commission can be set up as its standing committee for managing daily affairs. This Office can be set based on the existing CWRC, which shall be retained to execute the tasks assigned by the MWR.

Adequate Empowerment of the River Basin Management Authority

With construction and perfection of the socialist legal system in China, it has been made possible for this authority to manage the river basin on basis of law. The law-based management of river basin presupposes the establishment of sound laws and regulations.

Relations of various kinds in river basin management shall be specified by legislation, and the management authority shall be adequately empowered for it to carry out its responsibilities. A counterpart law–execution system for water administration shall also be established, to create a good water resources management environment governed by laws and regulations.

Establishment of a More Perfect Legal System For River Basin Management

The Water Act, the basic law for water resources management, shall be taken as the basis for such a legal system. Meanwhile, the actual conditions in the valley shall be given full consideration.

Therefore, when making amendments to the Water Act, it is proposed to highlight such points as follow. It shall be specified that the river basin management authority is the main body for river basin management.

Responsibilities and rights of river basin management and regional management shall be definitely described. The river basin management structure shall be tidied up. Functions of this authority shall be further consolidated and strengthened. Stronger policy support shall be given to this authority for its continual growth.

In addition, the enactment of River Basin Management Act shall be speeded up and the making of Changjiang Act shall be listed on the lawmaking plan. The proposed Changjiang Act will be the basic law for integrated water resources management in the valley.

The basic principle for river basin management shall be stressed. And such issues shall be clearly specified as long-range river basin planning, flood control, water administration and water resources development, water and soil conservation, water resources protection, hydrometry, river channel management, dyke construction, use of shoals, construction and management of flood diversion and detention basins, administration and management of backbone water projects in the valley and so on.

In the meantime, counterpart local and basin wide laws shall also be formulated, with a view to form a unified and complete legal system composing of national, basin wide and local laws for the proposed Changjiang Valley Water Commission to carry out its duties. An independent and adequately empowered law-execution body shall be created to execute the law for water administration.

Empowerment of the River Basin Management Authority

The proposed CVWC is the highest level decision-making and administration body for water management in the valley. All the important issues, except for those under the State macroscopical control, shall be decided by this commission on periodic or called at need meetings. The CVWC Office shall assume the responsibility for implementation and supervision of these decisions.

The proposed CVWC shall have the following primary functions: decision–making on key issues such as flood control, water and soil conservation and water pollution prevention; decision–making for water administration and for comprehensive development planning; proposal approval, examination or prequalification of key water projects; decision–making for comprehensive development investment plan, budgetary and final accounting; examination and approval of water allocation schemes; quality monitoring, progress supervision and appraisal and check of comprehensive development programmes; coordination and resolving inter-province water affair disputes; and administration and management of backbone projects on the main stem of Changjiang and its primary tributaries on behalf of the State investor. Water resources management practices at home and abroad have proven that integrated management on a basin basis is an efficient means for water management. As a river basin management agency accredited by the State, the CWRC has made substantial contributions to promoting water resources integrated management in the valley. With social and economic development, the needs for a new water resources management system oriented to the 21st century have become apparent.

The establishment of such a new system will guarantee the sustainable development of water resources in the valley. This new system shall have the following basic components: a powerful river basin management authority; more perfect and complete function system and legal system for river basin management; an independent adequately empowered law-execution body for water administration; sufficient power granted through legislation to this authority to fulfil its responsibilities; and policy support provided for the growth of the authority

NONPARAMETRIC APPROACH FOR PRESERVING INTERANNUAL

INTRODUCTION

An important goal of stochastic hydrology is to generate synthetic streamflow sequences that are statistically similar to the observed flow record. These sequences serve as inputs for Monte Carlo simulation of a reservoir system, to help identify plans and policies for efficient management of available water resources. A key requirement in stochastic streamflow simulation is that the generated sequences be "similar" to the observed flows. This implies that the distributional and dependence attributes of observed flows should be accurately reproduced in the simulations. The representation of seasonal to inter-annual dependencies commonly associated with sustained droughts or periods of large flows is of particular importance for reservoir system management. Absence of such dependencies in simulations can result in an inaccurate representation of the flows that are likely to occur. This can in-turn lead to biased reservoir operating policies causing both loss of revenue in reservoir operation, and a possible hazard for users downstream.

This paper presents an approach for stochastic simulation of seasonal streamflow sequences that attempts to reproduce such longer term dependence characteristics and the observed distributional attributes in the generated flow sequences. The approach is developed within a nonparametric density estimation framework that ensures accurate representation of the distributional attributes present in the historical flow record.

Stochastic simulation of seasonal flows has traditionally been approached using two different perspectives. Autoregressive moving average (ARMA) models have been commonly used to model both seasonal and annual streamflow sequences.

These models assume that the current flow is linearly related to previous observations. Many a times the actual flow values need to be transformed to an alternate variable that conforms well with the assumptions of linearity (or a Gaussian probability density) implicit in the model structure.

Use of such a framework offers an accurate representation of the dependence between the current and a few past flow values, but does not necessarily ensure that longer-term (seasonal to interannual) dependencies are accurately reproduced.

ANext, the generated annual or aggregate flow for each year is disaggregated or divided into the various seasonal components. This ensures that if the annual flow corresponds to a low flow year, the associated seasonal flows will also represent the same. While this offers a reasonable alternative to ARMA models, and also ensures that some measure of inter–

annual dependence is translated to the seasonal flow simulations, the resulting flow sequences offer only an approximate representation of the processes observed in the historical flow record. Proposed here is a seasonal streamflow generation approach that is free from several inherent disadvantages in ARMA models. Generated sequences reproduce both the short term as well as interannual dependence present in the historical flows.

Use of the nonparametric framework ensures that dependence and distributional attributes in generated flows are similar to those in the historical record. What follows is a brief background on nonparametric methods, their applications in hydrology and water resources, and how they can be used to formulate conditional streamflow simulation models.

The model is next applied to 105 years of Burrendong dam inflows on the Maquarie River in eastern NSW, Australia. We conclude with a discussion of the approach, its pros and cons, and mention some of the work that lies ahead.

NONPARAMETRIC APPLICATIONS FOR STOCHASTIC STREAMFLOW GENERATION

The past few years have seen a surge in applications of nonparametric methods for probability density and regression function estimation to a range of hydrologic problems.

Some of the applications related to the present work are–a synthetic streamflow resampling approach using nearest neighbour density estimation principles; a nonparametric alternative to the Autoregressive order p model (the NPp or the nonparametric order p streamflow simulation model); and, a nonparametric alternative to traditional disaggregation approaches (the NPD or the nonparametric disaggregation model).

Streamflow simulation is an exercise in conditional probability distributions. Simulation of flow X_t conditional to p prior flows ($X_{t-1}, X_{t-2}, \ldots, X_{t-p}$) involves estimation of the conditional probability density function $f(X_t \mid X_{t-1}, X_{t-2}, \ldots, X_{t-p})$. Similarly, disaggregation of an aggregate flow $Z = X_1 + X_2 +$

... + X_d into the d seasonal components (X_1, X_2, ..., X_d) requires estimation of the conditional multivariate probability density f(X_1, X_2, ..., X_d | Z). Conventional approaches assume certain distributional forms for the joint and marginal probability densities of the flow variables, from which the above conditional probability density functions are derived. These conditional densities are then expressed using parameters such as the mean, variance and skewness, and measures of dependence such as correlation.

As these methods rely solely on parameters (mean, variance, skewness, correlation) of the data to characterise the assumed probability density functions, they are termed parametric. Such methods are useful only if the assumptions about the underlying distributional forms are accurate.

One often comes across streamflow records that are not easily characterisable by the commonly used probability distributions. Nonparametric methods offer an efficient alternative to traditional parametric approaches. A nonparametric kernel probability density estimate is obtained by considering the cumulative effect of smooth functions called kernels placed over each sample data point.

Using a Gaussian kernel function, the multivariate kernel probability density $\hat{f}_X(x)$ of a d–dimensional variable set x at coordinate location x is estimated as:

$$\hat{f}_X(x) = \frac{1}{n} \sum_{i-1}^{n} \frac{1}{(2\pi)^{d/2} \lambda^d \det(S)^{1/2}} \exp\left(-\frac{(x - x_i)^T S^{-1} (x - x_i)}{2\lambda^2} \right)$$

where:

- x_i is the i'th multivariate data point, for a sample of size n,
- S is the sample covariance of the variable set x, and,
- λ is a smoothing parameter, known as the "bandwidth" of the kernel density estimate

The bandwidth, λ, is the key to an accurate estimate of the probability density. A large value of λ results in an oversmoothed probability density, with subdued modes and over-enhanced tails. A low value, on the other hand, can lead to density estimates overly influenced by individual data points, with noticeable bumps in the tails of the probability density. Several operational rules for choosing optimal values of the bandwidth λ are available in the lite.

PROPOSED APPROACH

This approach is aimed at accurately representing interannual dependence in simulated flows. Consider the flow at time t to be X_t, where t could represent annual, seasonal or monthly time steps. For example, for monthly flows, $X_1, X_2, \ldots, X_{12}$ would be the flows for the first 12 months, $X_{13}, \ldots, X_{24}$ the flows for the next 12 months, and so on.

The aggregate flow variable Z_t can then be defined as:

$$Z_t = \sum_{j-1}^{m} X_{t-j}$$

where m is the number of prior flows included in the aggregate variable. This study uses monthly flows and an annual aggregate level (m=12) to formulate the simulation model. The variable Z_t thus represents the annual flow during the past 12 months for the month being simulated, and its use as a conditioning variable enables proper representation of interannual dependence features.

Simulation proceeds from the following conditional probability density:

$$f\left(X_t \middle| X_{t-1}, X_{t-2}, \ldots, X_{t-p}, Z_t\right) = \frac{f\left(X_t, X_{t-1}, X_{t-2}, \ldots, X_{t-p}, Z_t\right)}{\int f\left(X_t, X_{t-1}, X_{t-2}, \ldots, X_{t-p}, Z_t\right) dX_t}$$

$$= \frac{f\left(X_t, X_{t-1}, X_{t-2}, \ldots, X_{t-p}, Z_t\right)}{f_m\left(X_{t-1}, X_{t-2}, \ldots, X_{t-p}, Z_t\right)}$$

where $f_m(.)$ represent the marginal probability density of the variable set. Note that the above conditional probability density is a function of (p+1) variables: Z_t and $(X_{t-1}, X_{t-2}, \ldots, X_{t-p})$. While use of the variables $(X_{t-1}, X_{t-2}, \ldots, X_{t-p})$ enforces a short term (till lag p) dependence structure in the simulated flow value, the aggregate variable Zt ensures that the annual dependence pattern is correctly represented. Also note that the conditional probability density in (3) has been specified as a function of p prior lags of X_t.

One needs to estimate the appropriate value for p in case of a real application using an order selection scheme such as the Akaike Information Criterion (AIC) [Akaike, 1974] or Generalised Cross Validation (GCV) . The authors recommend the use of GCV for estimation of the optimal model lag. The present application assumes p equal to 1 for the sake of simplicity.

The conditional density used for simulation then becomes:

$$f(X_t|X_{t-1},Z_t)=\frac{f(X_t,X_{t-1},Z_t)}{f_m(X_{t-1},Z_t)}$$

Using the kernel density estimator in, the conditional density in is estimated as:

$$\hat{f}(X_t|X_{t-1},Z_t)=\sum_{i-1}^{n}\frac{1}{\sqrt{2\pi\lambda^2 S^\wedge}}W_i\exp\left(-\frac{(X_t-b_i)^2}{2\lambda^2 S^\wedge}\right)$$

where: $\hat{f}(X_t|X_{t-1},Z_t)$ is the conditional probability density estimate;

S′ is a measure of spread of the conditional probability density, expressed as:

$$S^\wedge=S_{11}-\begin{bmatrix}S_{12}\\S_{1x}\end{bmatrix}^T\begin{bmatrix}S_{22}\,S_{2x}\\S_{2x}\,S_{xx}\end{bmatrix}^{-1}\begin{bmatrix}S_{12}\\S_{1x}\end{bmatrix}$$

where the covariance matrix of the variable set (X_t, X_{t-1}, Z_t) is written as:

$$Cov(X_t,X_{t-1},Z_t)=\begin{bmatrix}S_{11} & S_{12} & S_{1x}\\S_{12} & S_{22} & S_{2x}\\S_{1x} & S_{2x} & S_{xx}\end{bmatrix}$$

w_i is the weight associated with each kernel that constitutes the conditional probability density:

$$W_i = \frac{\exp\left(-\frac{1}{2\lambda^2}\begin{bmatrix}(X_{t-1}-X_{i-1})\\(Z_t-Z_i)\end{bmatrix}^T\begin{bmatrix}S_{22} & S_{2x}\\S_{2x} & S_{xx}\end{bmatrix}^{-1}\begin{bmatrix}(X_{t-1}-X_{i-1})\\(Z_t-Z_i)\end{bmatrix}\right)}{\sum_{j-i}^{n}\exp\left(-\frac{1}{2\lambda^2}\begin{bmatrix}(X_{t-1}-X_{i-1})\\(Z_t-Z_i)\end{bmatrix}^T\begin{bmatrix}S_{22} & S_{2x}\\S_{2x} & S_{xx}\end{bmatrix}^{-1}\begin{bmatrix}(X_{t-1}-X_{i-1})\\(Z_t-Z_i)\end{bmatrix}\right)}$$

bi is the conditional mean associated with each kernel:

$$b_i = x_i - \begin{bmatrix}S_{12}\\S_{1x}\end{bmatrix}^T\begin{bmatrix}S_{22} & S_{2x}\\S_{2x} & S_{xx}\end{bmatrix}^{-1}\begin{bmatrix}(X_{t-1}-X_{i-1})\\(z_t-Z_i)\end{bmatrix}$$

x_i and z_i represent observations, z_i being estimated using the 12 prior flows as expressed in equation. The conditional probability density estimate in can be viewed as consisting of n kernels having relative areas equal to weight wi, centered at bi, and having a spread proportional to S'.

Each of these are slices of the trivariate kernels that constitute the joint probability density of (X_t, X_{t-1}, Z_t), along the conditioning plane specified by (X_{t-1}, Z_t). The weight w_i depends directly on how far the kernel is from the conditioning plane.

A smaller weight implies that the kernel is far from the conditioning plane and does not make up a significant proportion of the conditional density estimate. On the other hand, a large wi implies that kernel i is close to the conditioning plane and constitutes a significant portion of the conditional density estimate.

Readers should note that the model proposed here is similar to the NP1 model of, except that the proposed model uses an aggregate flow variable in addition to the previous month's flow as the two model predictors. The use of the aggregate flow variable is to impose a longer term dependence in the simulated flows. Such dependence is missing in the NP1 or any other Markov order 1 dependence models.

APPLICATION TO BURRENDONG DAM INFLOWS, NEW SOUTH WALES, AUSTRALIA

The nonparametric simulation model was next applied to 105 years of reservoir inflows to the Burrendong dam in eastern NSW, Australia. The Burrendong dam is located on the Maquarie River and has an approximate catchment area of 7500 km^2. While flow data has been measured since the opening of the dam in 1967, the earlier periods of record have been estimated by the New South Wales Department of Land and Water Conservation using the observed rainfall record and a rainfall-run-off model.

This streamflow data poses many problems to the stochastic modeller. Firstly, there are several instances where the flow has stayed at fairly low levels for 6-10 months at a stretch. Secondly, there are several 'zeroes' in the flow record, which always pose a few challenges when prescribing a continuous probability density function. And lastly, this river is known to be susceptible to prolonged droughts, leading to long periods of very low flows (the minimum 11 month and 12–month average flows are respectively 0.8% and 2.2% of the mean annual flow).

One hundred realisations each 105 years long were simulated using the two nonparametric stochastic streamflow generation models. Statistics such as the mean, standard deviation, lag correlations and the coefficient of skewness, were computed for each month, and were found to be in good agreement with the historical values. This is to be expected given the structure of the nonparametric model. Results are not presented for lack of space. A comparison of some of the statistics of the annual flow volumes (summation of monthly flows over the water year) is presented in Table.

While both models are able to model the annual mean flow reasonably well, the NP1 simulations are unable to reproduce the observed annual flow standard deviation, coefficient of skewness, or lag–1 correlation. This is an important result that illustrates the ability of the NPL1 model to ensure that simulated flow values preserve distributional and dependence attributes at both monthly and annual scales. Figure 2.16 illustrates the reservoir storages estimated based on NP1 and NPL1 model simulations using the sequent peak algorithm. The NPL1 model simulations lead to reservoir storage volumes that are close to those estimated based on the historical flow record.

Table. Comparison of Observed and Simulated Burrendong Dam Annual Inflow Statistics

Statistic	Observed		NP1	NPL1
Mean (ML)	1096697	25th %ile	1,049,526	953,064
		Median	1,086,588	1,022,806
		75th %ile	1,159,303	1,102,851
Standard Deviation (ML)	1130626	25th %ile	830305	881,642
		Median	928528	995,120
		75th %ile	1007665	1,170,716
Skewness	3.01	25th %ile	1.56	2.18
		Median	1.91	2.91
		75th %ile	2.13	3.31
Lag–1 Correlation	0.114	25th %ile	–0.040	0.049
		Median	0.015	0.090
		75th %ile	0.078	0.134

NP1 model simulations, however, lead to substantially smaller storage volumes than those suggested by the historical flows. These smaller storage volumes are due to the lack of a longer–term dependence structure in the NP1 model simulations, and have obvious implications for any water management applications they might be used for.

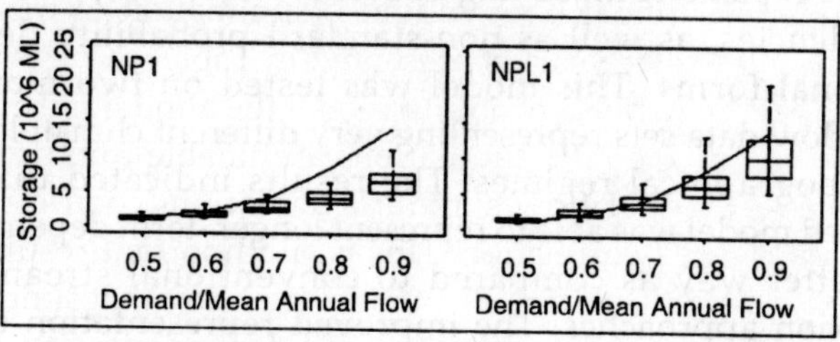

Fig. 2.16 Burrendong Dam Reservoir Storage Volume Estimates for Meeting Seasonally Non-Varying Demands

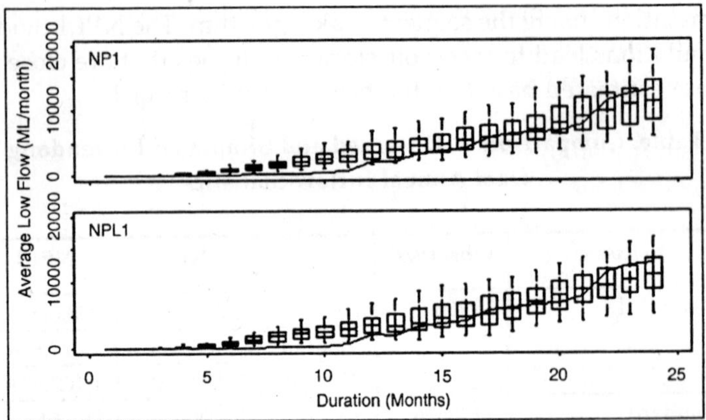

Fig. 2.17 Variation of Observed and Simulated Low Flow Values as a Function of Duration for the Burrendong Dam Monthly Inflow Record

Both models were also tested for their ability to simulate flows that would be likely in the event of a sustained drought. Figure 2.17 presents the lowest monthly flow rate as a function of duration. It is interesting that both models are able to properly simulate observed low-flow sequences for durations longer than 11 months. Neither model performs as well at simulating the worst 11–month drought on record. It is possible that more than one short term dependence variable is necessary to represent such severe flow conditions.

SUMMARY

A synthetic streamflow generation model was presented that was capable of modelling both short term and interannual dependencies, as well as non-standard probability density functional forms. This model was tested on two monthly streamflow data sets representing very different climatological and topographical regimes. The results indicated that the proposed model was able to represent longer–term dependence in a better way as compared to conventional streamflow simulation approaches. The improved representation of the longer-term dependence lead to significant improvements in the representation of reservoir storage volumes. It should be noted that the results presented here were based on arbitrary choices for the number and type of variables used to represent the short and long-term dependence. It is likely that results will improve further if these variables are chosen based on an elaborate sensitivity analysis. Efforts are currently underway to develop a method for selecting model variables that impart the short and long-term dependence in an optimal manner. Efforts are also underway to extend the proposed method for use with flow records at multiple locations.

3

Using Treated Wastewater to Recharge Reclaimed Land

INTRODUCTION

In the coming years, land reclamation works at the eastern coastal area of Singapore would yield a huge reclaimed land area of over 2500 hectares. The land is being reclaimed by filling up the original seabed with graded gravel sand to depths exceeding 10m.

This piece of land may be considered as an unconfined coastal aquifer with the possibility to be used for groundwater storage. With a centralised wastewater treatment plant to be built on a part of the reclaimed land, the possibility of using the treated wastewater to recharge this aquifer is attractive. A rough estimate shows that the water thus recharged, and recovered subsequently, could be up to one third of the total current water demand of Singapore.

The major challenges facing such a scheme are saltwater intrusion due to excessive pumping from the aquifer and the quality issues of the treated wastewater to be injected and subsequently extracted from the aquifer.

The former problem, which is physical in nature, can be resolved by carrying out a detailed analysis based on groundwater modelling and proposing an optimum injection-extraction well system to avoid saltwater intrusion into the aquifer.

The later problem is however more complex in nature and needs considerable amount of research on the water quality transformation occurring through the aquifer, estimation of the rate and extent of clogging of the aquifer medium and the measures to rectify these problems. With the scarcity of literature on this topic, the job to resolve water quality issues has become more challenging. A significant large scale field experiment on solute transport has been reported by Mackay et alwhere they have considered the transport of synthetic organic chemicals through the aquifer. Another study similar in nature to the former one has been reported by Ding *et al.*

In the present study, a laboratory scale model is being utilised to observe the changes occurring in the water quality as the treated wastewater passes through the sand brought from the reclaimed site. The treated wastewater is initially made to pass through a filter made from the reclaimed sand in order to observe the clogging phenomenon.

EXPERIMENTAL SET-UP

The sand was brought from the reclaimed site and filled in a flume as shown in Figure 3.1. The straight portion in the middle of the flume may be considered as the test section and the flow in this part may be reasonably assumed as two-dimensional. Two PVC pipes of 300mm diameter, with perforations and well screen on their bottom portions, are installed at both the ends of the flume as 'recharge' and 'extraction' pipes.

Piezometers are installed along the centreline of the flume at 30cm c/c. Those are 5cm diameter PVC pipes with perforations and well screen wrapped on the bottom portion. A typical piezometer is shown in Figure 3.1.

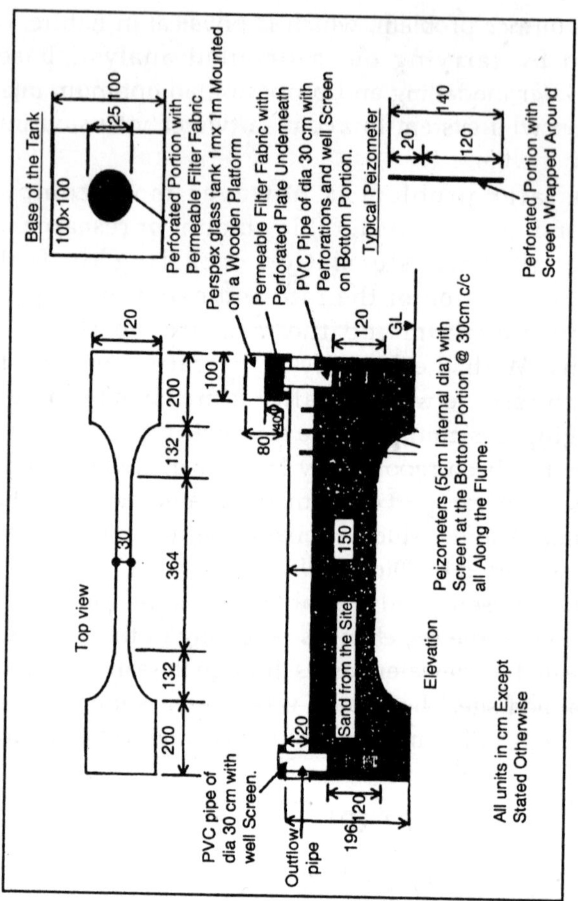

Fig. 3.1 Schematic Diagram of the Experimental Set-up

A filter is placed at inlet in order to monitor the possible change in the permeability of the sand with the passage of time due to clogging by the wastewater constituents. The filter is fitted with two piezometers in order to observe the head difference between its inlet and outlet, to allow the permeability coefficient to be monitored.

Physical Properties of the Sand

The source of the sand is ocean bottom near Malaysia

and Indonesia. The chemical analysis would be done in order to ascertain the chemical composition of the filling material. The results of the sieve analysis of a sample taken from the reclaimed site are presented in Table and the gradation curve is shown in Figure 3.2. The soil may be described as loose light brown gravelly sand.

Table. The Result of Sieve Analysis for the Sand Used in the Experiment.

Sieve Size (mm)	% Finer	Classification	
		Soil	%
3.35	100	Gravel	13.2
2.00	86.8	Sand	86.8
1.18	63.0		
0.60	32.9		
0.425	16.5		
0.30	8.2		
0.212	2.8		
0.15	0.6		
0.063	0.0		

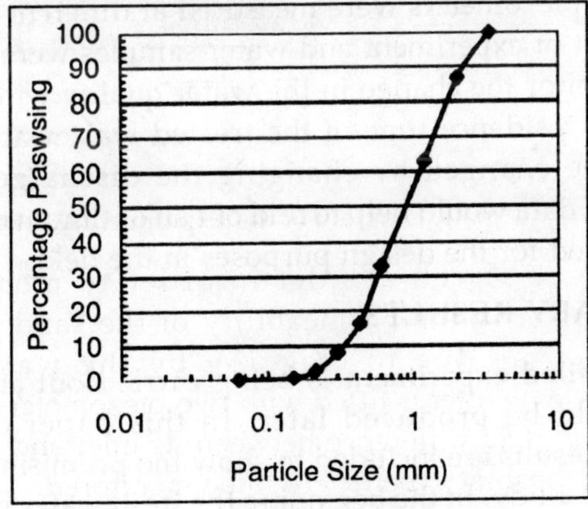

Fig. 3.2 Gradation Curve for the Sand tThat was Brought from the Reclaimed Site to be Used in the Experiment

Quality Parameters of the Wastewater Used in the Experiment

The treated wastewater to be used in the present study was brought in from a wastewater treatment plant nearby. It was tested at the Environmental Laboratory at Nanyang Technological University, Singapore and some of the important constituents would be shown later.

METHODOLOGY

The present experiment is designed to gather information regarding the water quality transformation through the sand and the changes occurring in the sand due to the passage of treated wastewater. In order to carry out an experiment that matches the real phenomenon in the field, the whole flume was initially saturated with fresh water. Some preliminary tests were carried out to ascertain the hydraulic conductivity of the sand by measuring the discharge (volumetric method) and hydraulic gradient (from the piezometers) under steady conditions.

The sand filter was then installed at the inlet and the treated wastewater was then pumped through the filter to the flume at a constant discharge. The steady state was achieved by adjusting the valve opening at the bottom part of the outlet. The water levels in the piezometers were measured at different intervals after the start of experiment and water samples were taken in order to monitor the change in the water quality with respect to time. The residence time of the treated wastewater in the flume can be changed by changing the discharge, so the experimental data would help to find out an optimum residence time to be used for the design purposes in the field.

PRELIMINARY RESULTS

The detailed experiment is being carried out and those results would be produced later. In this paper only the preliminary results are included to show the promising nature of the present study. In the beginning the fresh water was used to saturate the sand in the flume and a discharge of 0.018 l/s was maintained, a value obtained by trial and error to achieve steady state flow through the soil medium.

Table. Water Quality Parameters at Various Points Along the Flume (Distance X is in Meters)

Parameter/ Sample	TDS (ppm)	TSS (mg/L)	Calcium	Sodium	COD (mg/L)	Colour	pH	Conduc -tivity (ms/cm)	Turbi- dity (NTU)	Alkali- nity (mg/L)
Before filter	1475	50.59	53.11	252.48	388.00	87.80	7.430	1.93	10.00	79
After filter	1505	0.0095	60.20	252.65	275.00	38.76	8.03	2.02	1.06	106
P20 (X=0.00)	990	0.00	51.68	215.33	367.00	37.90	7.41	1.35	3.21	105
P18 (X=0.68)	1328	0.00	58.57	252.35	366.00	19.52	7.91	1.61	1.06	100
P15 (X=1.54)	1212	0.00	66.32	252.58	115.00	14.07	7.88	1.43	0.49	85
P12 (X=2.43)	854	0.00	80.22	133.63	156.00	7.14	7.14	1.24	1.55	66
P09 (X=3.27)	538	0.00	66.52	58.81	66.00	3.01	8.00	0.78	2.60	62
P06 (X=4.13)	325	0.00	40.57	29.40	54.00	2.15	8.13	0.50	1.04	80
Parameter/ Sample	TDS (ppm)	TSS (mg/L)	Calcium	Sodium	COD (mg/L)	Colour	pH	Conduc -tivity (ms/cm)	Turbi- dity (NTU)	Alkali- nity (mg/L)
P03 (X=5.01)	311	0.00	40.35	30.95	57.00	2.46	8.21	0.46	1.26	99
Outlet (X=6.15)	311	0.00	37.05	252.65	56.00	3.54	8.17	0.47	0.70	93

The hydraulic gradient was measured from the piezometers in the flume and the hydraulic conductivity was calculated by using Darcy's law, which had a range from 64 to 67 m/day for various readings taken during three days. In the next step the treated wastewater was made to flow through the filter then allowed to enter in the flume at inlet. A significant improvement was observed in the water quality by virtue of most of the parameters tested in the samples. The results for the samples taken one day after the test started are shown in Table. It may be observed from Table that there is a significant (approximately 80%) reduction in Total Dissolved Solids (TDS) when the treated wastewater is allowed to flow through the sand. It may be also be seen that the filter does not affect TDS, it completely eliminates the Total Suspended Solids (TSS), though. A 76% reduction in the Conductivity, 85% reduction in Chemical Oxygen Demand and 90% improvement in colour are also obvious. Calcium, Sodium, pH, turbidity and alkalinity show random values, the reason for which is under investigation at present.

The main contribution of the filter is in the reduction of TSS, COD, colour and turbidity where a significant improvement is observed after the filter. A preliminary laboratory study on the recharge of reclaimed land with treated wastewater has been presented. The initial results show that the experimental set-up used in the present study would produce promising data that would be helpful for the design of a large-scale storage and recovery experiment on the actual reclaimed land in the coastal area of Singapore.

PROFILE OF MULTIPLICATION FREQUENCY CURVE TYPE III (X-III)

The multiplication theorem of probability usually is neglected in the theoretical distribution of flood or frequency curve of flood, which often applies the mean value that has neither resistance nor robustness. Considered that the flood is a random variable, it should follow the multiplication theorem of probability. Meanwhile, the median has the different attribute from the mean value, it is both resistance and robustness.Based the multiplication theorem of probability and the attribute of the median, this paper briefly introduces the theory, the characteristic, the application method and example about the multiplication frequency curve type III(X–III for short).This simple probability distribution is available for the flood occurred in the kinds of rivers and streams.

THEORY

Based the multiplication theorem of probability and the attribute of the median, and assumed D=(a–1)/(a^(k^c)–1), then the X–III expresses as follows:

The probability distribution function:

$$P(K>k)=1-0.5^D$$

The probability density function:

$$f(k)=-\ln 0.5 * 0.5^D * (a-1) * a^{(k^c)} * c * \ln a * k^{(C-1)}/(a^{(k^c)}-1)^2$$

The inverse of the probability distribution function:

$$k=\{\ln [(a-1)*\ln 0.5/\ln (1-P)+1]/\ln a\}^{(1/c)}$$

Where k=X/X0, X0= median of the sample, P(k>1)=0.5 is frequency of median. A and c are two parameters of probability density function, which restrict to a>1,c>0.

CHARACTERISTICS

The Clearness and Lucidity of the Theory

The multiplication theorem of probability is as follows:

K=1(median)	P(K>1)=0.5^1	K=2	P(K>2)=0.5^2
K=0.1	P(K>0.1)=0.5^0.1	K=0	P(K>0)=0.5^0=1
K=∞	P(K>∞)=0.5^∞=0	K=k	P(K>k)=0.5^k (4)

The expressions names X–I, and is a single curve, Cv=0.999, Cs=1.98. It is neither flexibility nor adaptability. When two density parameters(a and c) of probability are added, this curve calls X–III as the expressions.It is not only flexibility but also adaptability. The expressions is the extension of the multiplication theorem of probability.

Resistant and Robust, Good Flexibility and Little Error

X–III is based on the multiplication theorem of probability, and the median represents location of distribution, so X–III is very resistant and robust.

Moreover, the probability density function as the expressions includes exponential function and power function, and can be effectively applied to the upper tail and lower tail distribution. Therefore, X–III has the good flexibility and the small error in the fitting curve.

The Simple Usage

The expressions, the inverse function of the probability distribution, is algebraic. Thus, known P, a and c, k can be calculated by using expressions as tables. When c=1,the curve calls X-III_1;when c=2, the curve calls X-III_2.

- X-III_1 is available for a instantaneous value such as a flood peak;
- X-III_2 is available for a non-instantaneous value, such as flood volume and precipitation.

A curve can be directly fitted by using a and c, not using Cv, Cs and Φ. The application method is very simple.

No J-Shaped Distribution and no Appearance k<0

Both tables show: when k<1,tail ofX-III is not level, namely not J-shaped distribution. And there is no appearance of k<0 which is the disadvantage of P–III.

METHODS OF FITTING CURVE

Preliminaries: Calculation of Median x0 and k=x/x0

After sorted samples, while n=odd, median X0 is the central value of sort order; while n=even, X0 is the mean value of two central value in the order. Thus, k=X/X0, empirical frequency P=m/(n+1).

Eye Estimate Curve Fitting

By selecting k value in table,the curve fitting process is

simple and easy by using parameters a and c, completely not using Φ,Cv and Cs.

Optimizing Curve Fitting

Considering the expression is algebraic, a optimizing process of fitting curve can be programmed in the computer. This method gets rid of the fitting trouble of using Φ,Cv and Cs.

OTHER USAGE: CALCULATION Cv AND Cs UNDER A KNOWN A AND C

Sampling more than 10000 data in X-III curve using the expressions (3) is necessary, thus n=10000, p=m/(n+1). Through a and c obtained using the methods-mentioned, k and X=k*X0 can be calculated. The Cv and Cs values can be got using these sampling data and the formula of matrix method.

EXAMPLE: FLOOD FREQUENCY ANALYSIS

The observed data from three hydrological stations is used inX-III, and doesn't include historical flood and paleo-flood; the flood data which includes historical flood and paleo-flood in P–III.The results of these two methods show the same effect.

The results verify: the analysis results of X–III frequency without paleo-floods are the same as the results of P–III with paleo-floods:

1. X–III is based on the multiplication theorem of probability and median attribute, resistibility and robust. The theory of probability statistical distribution of three parameters is clear.
2. X–III is available for the upper tail and lower tail distribution, its flexibility is wide, and can fit the distributions of kinds of rivers and streams.
3. X–III can be fitted direct by two density parameters a and c, and is simple. The Cv and Cs value can be calculated by using parameters a and c

TERRITORIAL INFORMATION SYSTEM FOR PLANNING

The proper programming of measures for complex

drainage networks relies on their characterization in the framework of a Territorial Information System capable of describing their functions and establishing the effects of any changes in management policies by means of suitable simulations. The fine-adjustment of such a system is particularly essential when, alongside the traditional hydraulic function of disposing of flood water, it also becomes necessary to use the networks for other purposes, such as to improve run-off water quality.

The development of an adequate Territorial Information System can prove complex, especially in the case of artificial drainage networks in which changes are frequently introduced, new channels are built and new plants are installed, all of which must be input into the system without interfering with its previous architecture. A system of this kind has been fine–adjusted for the drainage network of the Veneto lowland area, which is unquestionably one of the most complex in terms of extension and density of channels and flow–control and monitoring installations. This network, which is subdivided into 20 drainage districts, each managed by a Land Reclamation Consortium, extends over an area of approximately 1'184'000 hectares, 187'000 of which lies below mean sea level, 332'000 are mechanically drained and 95'000 are alternately mechanically drained.

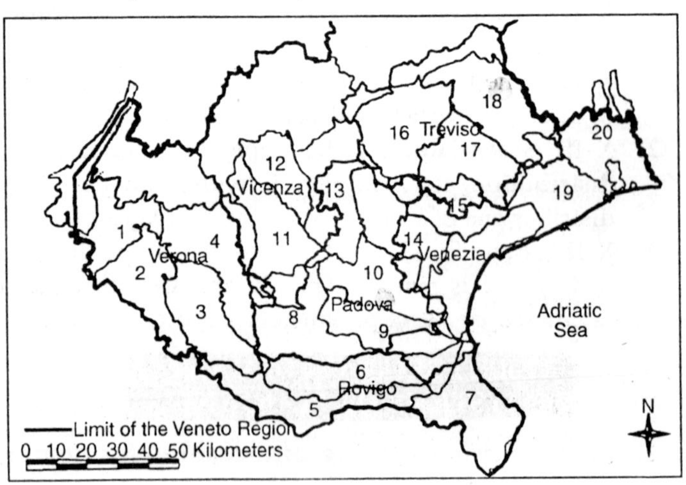

Code	Land Reciamation Consortium	Area (hectars)
1.	Adige Garda	54'859
2.	Agro Veronese Tartaro Tione	53'280
3.	Valli Grandi e Medio Veronese	61'563
4.	Zerpano Adige Gua	77'463
5.	Padana Polesana	60'848
6.	Polesine Adige Canalbianco	66'481
7.	Delta Po Adige	66'298
8.	Euganeo	70'662
9.	Adige Bacchiglione	49'037
10.	Bacchiglione Brenta	58'247
11.	Riviera Berica	57'216
12.	Medio Astico Bacchiglione	38'320
13.	Pedemontano Brenta	70'662
14.	Sinistra Medio Brenta	56'731
15.	Dese Sile	43'464
16.	Pedemontano Brentella di Pederobba	64'699
17.	Destra Plave	52'995
18.	Pedemontano Sinistra Piave	72'700
19.	Basso Piave	56'004
20.	Pianura Veneta Livenza Tagliamento	55'913

Fig. 3.3 Drainage Districts of the Veneto Region Involved in the Study and Their Corresponding Surface Areas.

THE STRUCTURE OF THE TERRITORIAL INFORMATION SYSTEM

A Territorial Information System represents a fundamental tool for managing and programming measures that a territorial organization such as the Land Reclamation Consortium is required to implement.

Within the system, data of various kinds can be organized and integrated, alongside programmes specifically designed for their analysis and processing. The result is a somewhat complex structure, composed of the following main components: the information and data needed to define and deal with the problems to which the system is tailored; the management functions of these information, *e.g.* aggregations, updating, research and consulting; the programmes specializing in processing the information in a spatial and temporal sense, depending on the purposes for which the system has been conceived; the function of representing the information that has been input or derived from the processing operations.

The Territorial Information System is based on a GIS (Geographical Information System), which has the following particular features: the opportunity to manage large amounts of spatially-referenced data and tables of attributes referring to them; the availability of analytical methods that deal specifically with the geographical component of the data. the automatic topological construction of the representation, the superimposing of themes and areas, the network analysis, the application of spatial models; the organized handling and feedback of large quantities of data, in both vectorial and raster format.

The GIS consequently enables the creation of an environment in which all the details describing the territory can be placed in relation to each other so as to generate further information. Hence the considerable benefits that the GIS offers to the Territorial Information System, *i.e.* the assessment of interactions between data sets, the prediction in real time of the influence and impact of any changes being planned, support for decision–making processes, the optimized management of the territory and of any activities that are performed in the area.

The description of the physical entity representing the context of the activities of the Land Reclamation Consortium, such as the hydrographic basins, channels and pumping stations, is implemented in the abstract form of area, stretch or arc, and node; the connection between these entities is guaranteed by a common code that enables a shift from one to the other type of information on the basis of their correlations.

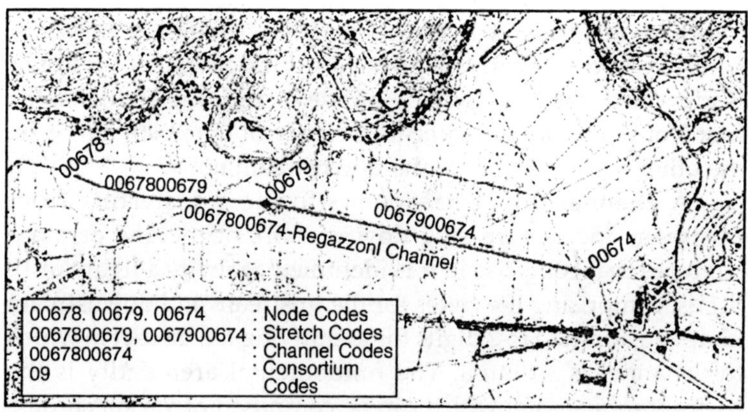

Fig. 3.4 Example of the Codification of Nodes and Stretches.

The nodes identify any geographically–referenced point of information that has a particular significance from the physical point of view, for example: the source or origin of a water course; the confluence of two or more water courses; any hydraulic structures such as siphons, flow–control apparatus, pumping stations, check gates, outlet works, side spillways; stations for measuring water quality, rainfall, temperature, air humidity, solar radiation, water level and flow rate.

The nodes have been unequivocally codified by attributing to each of them a number composed of 7 digits: the first 2 represent the code number identifying the Land Reclamation Consortium, and therefore identify the district to which the node belongs; the other 5 digits characterize the number of the node within the Land Reclamation Consortium. This number is always sufficient to identify all the nodes and, at the same time, it poses no hindrance to the introduction of new elements for a further definition of the drainage networks.

The set composed by the Land Reclamation Consortium code plus the node code represents a single code. Each node has been associated with an information file, which can be implemented with or without the addition of descriptive information. The stretch coincides with a sequence of lines that joins two nodes; it corresponds to a stretch of water course and is identified unequivocally by a code comprising 10 digits, the first 5 of which are the same as the code of the upstream node, the other 5 for the code of the downstream node.

A channel thus corresponds to the set, starting from upstream and progressing downstream, of the one or more stretches that comprise it: it is identified again by a figure of 10 digits, comprising the codes for the upstream and downstream nodes, in addition to 2 digits for identifying the district in which the channel is situated. The fundamental area entity is the hydrographic basin, which can be composed of various orders of sub-basin.

The basin entities are distinguished by a basin code and, where applicable, by a sub–basin code, which is also included in the channel files, so that the passage between a channel and the basin into which it flows is immediately clear. The hydrographic basin has been associated with a suitably-structured information file.

THE HYDROGRAPHIC AND STRUCTURAL ELEMENTS IMPLEMENTED

The research was carried out with reference to the 20 drainage districts of the Veneto Region; to each of these districts was attributed an identification code, which is consistent with the one attributed by the Veneto Regional Authority at the time of their establishment.

The geographical data were implemented using the elements on the Regional Technical Map, on a scale of 1:10,000, suitably raster-scanned and geographically-referenced, and thus on a scale of considerable detail for establishing the course of the channels and the position and functions of the various drainage plants.

The raster–scanned database provided with all the needed cartographic support from the beginning of the study, since the numerical format of the above-mentioned map was not available for the whole Region.

Table. Length of the Hydrographic Network, Number of Nodes for Identifying the Stretches of Water Course and Number of Installations.

Code	Land Reclamation Consortium	Hydrograph-ic Network [km]	Mean Network Density [km/km²]	No. of Nodes	No. of Installations
01	Adige Garda	177	0.984	193	0
02	Agro Veronese Tartaro Tione	294	0.907	369	53
03	Valli Grandi e Medio Veronese	1271	2.019	1'146	336
04	Zerpano Adige Guà	1109	1.475	1'361	225
05	Padana Polesana	831	1.462	926	291
06	Polesine Adige Canalbianco	696	1.080	630	233
07	Delta Po Adige	540	1.035	929	121
08	Euganeo	1'004	1.485	1'410	666
09	Adige Bacchiglione	666	1.358	709	385
10	Bacchiglione Brenta	878	1.507	937	449

Code	Land Reclamation Consortium	Hydrograph-ic Network [km]	Mean Network Density [km/km²]	No. of Nodes	No. of Installations
11	Riviera Berica	720	1.252	902	188
12	Medio Astico Bacchiglione	676	1.759	1'090	164
13	Pedemontano Brenta	1'711	2.421	1'986	809
14	Sinistra Medio Brenta	1'227	2.163	885	410
15	Dese Sile	601	1.383	402	175
16	Pedemontano Brentella di Pederobba	1'157	1.788	738	200
17	Destra Piave	638	1.204	431	151
18	Pedemontano Sinistra Piave	627	0.862	479	41
19	Basso Piave	639	1.141	548	112
20	Pianura Veneta tra Livenza e Tagliamento	969	1.733	2'310	378
	TOTAL		**1.384**		

The installations belonging to each Land Reclamation Consortium were described by means of suitably designed files, completed on the strength of known data and subsequently integrated with further information and updates. The number of nodes for identifying the stretches forming the network and the flow-control and monitoring installations considered are illustrated in Table.

This table gives an idea, first of all, of the considerable extension and density of the drainage network. The large number of nodes that had to be input to describe the network, amounting to approximately one node every 0.89 km, bears witness to its complexity and the degree of detail of the information that proved necessary in order to represent it.

In this context, the number of installations included in the Territorial Information System is remarkable, amounting to 5,387, mainly composed of pumping stations, check gates, flow-control and monitoring apparatus.

There are many channels of considerable size, and installations of major importance, particularly pumping stations, which have been built on them. In total there are 316 pumping stations, with a pumping capacity of approximately 1,400 m3/s.

PECULIAR FEATURES OF THE TERRITORIAL INFORMATION SYSTEM

The elements acquired on the structure of the drainage network and the corresponding installations contribute a good deal towards a better understanding of the relationships between hydrography and environment.

The proposed Territorial Information System enables this information to be processed within the geographical context to which it belongs, even in an administrative sense, facilitating the immediate identification of the number and features of the manifolds falling within a given municipality or province, and thus establishing the possible relations between them and the activities to be programmed in the territory.

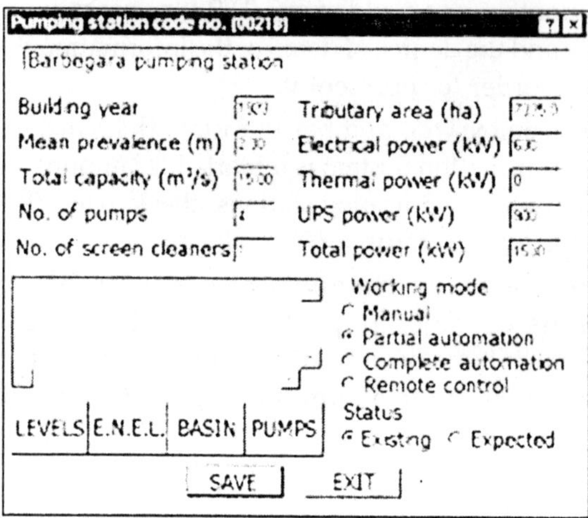

Fig. 3.5 View of a Pumping Station and Example of Some of the Data Implemented

Similarly, the associated files enable a detailed description of the installations, from which many kinds of information, such as water levels, types of equipment installed, or run-off flow rates can be extracted as necessary. The information of an alphanumerical nature can also be integrated with further details in graphic form, progressively implemented according to need and availability. For instance, Figure 3.5 shows a picture and some of the information implemented for a pumping station, Figure 3.6 shows the same for a check-gate. The geographical aggregations processed for each basin or drainage district can be composed to obtain data of hydraulic and functional nature relating to any geographical context: for example, within the area considered in the study, the subset comprising the basins discharging into the Venetian Lagoon is of particular interest. Of course, the potential applications of the Territorial Information System are numerous.

Fig. 3.6 View of a Check Gate and Example of Some of the Data Implemented.

First of all, it is worth mentioning its uses in a design context and, more in general, in the hydrological and hydraulic modeling sectors, which need the support of the data provided by the system about the structure of the drainage networks and the operation of the flow–control installations. In addition, the Territorial Information System can prove particularly valuable when it is used in relation to the development of environmental activities, characteristic of the work of the Land Reclamation Consortium alongside the traditional hydraulic purpose. In this context, it is particularly worth mentioning the drainage network management methods designed to reduce the concentration of pollutants in the run-off water. In the Venetian Lagoon watershed, for instance, these techniques are being proposed mainly to reduce the quantities of nutrients in the drainage networks flowing into the lagoon by means of various procedures based essentially on increasing the water retention time in the channels.

To be implemented properly, the programming of such measures and management techniques must be part of a global and detailed understanding of the territory that only a Territorial Information System of the kind discussed here can provide

HUMAN AND WILDLIFE HABITATS IN A NATURAL RESOURCES

The Mahaweli Ganga Development Scheme (MGDS) has been the largest multipurpose natural resources development project undertaken in Sri Lanka at least in the last five centuries. The sources of water for the MGDS have been the Mahaweli river, the largest and longest river in Sri Lanka and the adjacent rivers in the project area. There are thirteen independent development areas in the MGDS identified as System-A, System-B, System-C .... System-L and System–M. Of the thirteen Systems, H, B, C, G and L have either been fully or partly developed up to now, having implemented the project in the year 1970.

The principal proposals of the scheme, in view of the socio-economic problems faced by the country in the sixty decade were as appended below:

- Provision of gravity irrigation to nearly 360,000 Ha. of new and existing agricultural lands in the dry zone,

- Settlement of about 350,000 farm families,
- Generation of electrical energy from hydro-power through installed capacity of over 500 MW,
- Employment opportunities for over a million people in agriculture and related activities.

PLANNING CONCEPTS

The cardinal concept encompassing the MGDS has been the Man, his Habitat and his Economy within an integrated development system. The fundamental issues on which the entire scheme was based were, irrigated agriculture, human settlement and environmental conservation. The development of suitable lands for agriculture by means of gravity irrigation has been the principal concern in the planning.

Many, integrated development projects based on irrigation and human settlement cause environmental changes during and after development. When planning such projects, it is important to address environmental issues, which may arise due to various changes taking place in the different systems of the environment. Each system or sub–system can therefore, be represented by the Terrestrial, Aquatic and Human environments, to which all functions in a system are related.

The key resources, which are involved in these three environments are the water, land, flora, fauna and human.

ACCELERATED MAHAWELI GANGA DEVELOPMENT SCHEME

In the year 1977, the government decided to accelerate the implementation of the MGDS.

It was envisaged that the acceleration would include the following:

- Construction of storage reservoirs
- Provision of irrigation and social infrastructure
- Cultivation of 80,800 Ha. new lands and upgrading of 14,350 Ha. of existing lands in the Systems, designated as A,B,C and D.

The lowland extents which were to be brought under irrigated agriculture are shown tabulated below.

Reservoir	Surface area at FSL (Ha.)	Net storage capacity (MCM)	Irrigation Potential (Ha.)
Kotmale	970	408	20,000
Victoria	2270	690	53,000
Randenigala	2750	860	35,000
Rantembe	-	21	-
Ulhitiya/Rathkinda	2270	146	-
Maduru Oya	6280	430	16,000

TERRESTRIAL ENVIRONMENT

Climatic conditions in the project area are largely influenced by two annual monsoon seasons. From October to February, the area is generally under the NE monsoon. The SW monsoon prevails from March to September. The mean annual rainfall ranges from 1650 mm in the lowlands to 5300 mm in the mountains, where Kotmale, Victoria and Randenigala are situated. The annual temperature variations are generally less than 5°C from the average 27°C across the lowlands. Throughout the dry zone, the annual evapotranspiration rates normally exceed precipitation levels. The land uses of Systems A,B,C and D by the year 1979 are shown in the table below.

Development Area	Improved lands (Ha.)	New Lands (Ha.)	Other crop lands cultivated (Ha.)	Total cultivated (Ha.)
System - A	3,500	14,000	-	17,500
System – B	8,850	25,800	5,500	40,150
System – C	-	19,550	-	19,550
System - D	2,000	10,950	5,000	17,950
Total	14,350	70,300	10,500	95,150

AQUATIC ENVIRONMENT

Most of the small reservoirs in the project area could be categorized as highly productive or eutrophie systems, prior to augmentation with diverted waters. A number of villus (swamps) in the project area also support full time fishing activities. The flood–plains provide a valuable nursery grounds for many fish species. Fishing activities are well developed in the estuaries portions of both Maduru Oya and Mahaweli river. Villus (Swamps) are marshy areas which occur in association with the riverine flood-plain. There are more than 50 identified villus in the project area. The flood-plain of Mahaweli river occupies about 12,800 Ha. Villus are highly productive biological communities. Villu habitats are very favourable for many wildlife species and high concentrations of birds and mammals are usually observed here.

Mangrove swamps occur in the coastal areas, extending inland with tidal intrusion. However, mangrove swamps in the project area have been substantially reduced due to clearing of jungle. The aquatic weed species, which occur in abundance in the project area include water hyacinth, floating ferns (salvinia), water lettuce and cattails.

HUMAN ENVIRONMENT

The overall population in the project area was probably around 200,000 by year 1979 or so. In the year 1979 the total population in System-B was estimated at 25,515. There were four reported aboriginal villages. The total population in System-C in the year 1979 was 29,776. A rough approximation of population in both Systems A & D had been 30,000 and 100,000 respectively.

The land alienation policy adopted in the settlement planning and implementation in System-B and System-C under the MGDS has been as shown below:

- Size of homestead - 0.2 Ha.
- Size of lowland farm - 1.0 Ha.
- Horticulture lands–variable
- Pasture lands–variable

The size of a hamlet or a village in Systems B & C is different to that of Systems H, G or L and there are homesteads in the range 200 - 300 in each of Systems B & C.

WILDLIFE

The project area was home to a variety of wildlife habitats such as villus (swamps), grasslands and numerous forest types. Several designated wildlife reserves were partly or wholly situated in the project area. A Nature Reserve has to provide complete protection for wildlife while accommodating existing land rights. The project area inhabits diverse fauna and includes several endangered and two threatened animal species. These species are all found in other parts of Sri Lanka, and except for the toque macque, purple–faced langur and red-faced malkoha, are also distributed in other parts of the world. However, all nine species are generally considered to be threatened or endangered in their respective world distributions.

Of the endangered/threatened species, the Bengal monitor has the most extensive distribution within the project area. The red-faced malkoha is confined to the riverine forest in the northern part of System–C. The toque macaque and purple faced langur are most commonly found in riverine forest or near major reservoirs. Leopards were relatively numerous in northern part of System–C. Similarly, elephants were abundant in northern part of System–C and in the Somawathie National Park.

Of the estimated total of 2000 wild elephants remained in Sri Lanka, about 800 were presumed to inhabit the project area. In the dry season, elephants and other wildlife tend to congregate in the vicinity of perennial streams, reservoirs, swamps or in riverine forests. As surface waters rise with the onset of the monsoon rains, they disperse and enter upland areas where they graze throughout the wet season. The project area was also the natural habitat for a number of endemic animals who are found only in Sri Lanka.

These include eight species of fish, four amphibians, nineteen reptiles, eight birds and three mammals. Furthermore, 53 endemic plant species and 23 very rare plants have been identified as occurring in the project area. In addition, large mammal species such as wild boar, sloth bear, deer and sambhur were also observed in the project area. Also, majority of the 251 resident bird species found in Sri Lanka were observed in the project area. During the winter months elsewhere, about 75 more bird species migrate into the area from Europe and other countries in the region. Some field surveys conducted in the late 70's had indicated that a high level of wildlife activity, was identifiable for certain localities in the project area. There were categorized as "critical habitats". and included, Somawathie National Park, Mahaweli Flood Plain between Wasgomuwa and Somawathie National Park, Northern part of System–C adjacent to Wasgomuwa National Park and Wasgomuwa National Park.

ENVIRONMENTAL IMPACTS ON HUMAN SETTLEMENTS AND WILDLIFE

A significant social impact of the development exercise has been the transition from small isolated village, societies to production oriented large scale colonization schemes. The population of the project area was expected to rise drastically by one million people. New settlers are compelled to use guns to scare encroaching animals, causing loss to the wildlife. Extensive extents of wildlife habitats have been removed in clearing the forest and creating human habitats. Based upon previous records, a reasonable estimate of crop loss in project areas of this nature has been about 20% due to physical wildlife damage and feeding.

Due to clearing, a large portion of the forest in the project area was lost, resulting greater need for timber and fuel wood. There have been a loss of large numbers of animals and plants, many from the nine endangered and threatened species or more than 90 endemic species. The carrying capacities and population for most wildlife species have declined. The viability of the endangered species have been further reduced. The critical wildlife habitats of extents 5500 Ha. and 15,000 Ha. In Somawathie National Park and in system–C respectively have been eliminated. Breeding and feeding grounds for a number of fish species, waterfowl and migratory birds have been lost.

LAND ALLOCATION FOR HABITATS

Of the four systems A,B,C and D in the project area, only Systems B and C have been developed up to now, with the provision of irrigation and social infrastructure, facilitating irrigated agriculture and human settlement. It was envisaged with the implementation of integrated rural development, in System–B and System–C that 45,000 and 28,000 farm families would be inducted into the project area for irrigated agriculture. The irrigation and social infrastructure necessary for the new society of people would be provided replacing the natural forest and the wildlife. However, the development of new lands for cultivation by the settler-farmers had to proceed at the expense of natural habitats and the fauna and flora associated with them. The social values derived from cultural and religious practices and spiritual traditions in Sri Lanka demand the need to balance development goals with wildlife conservation. Therefore, it was considered to implement the most feasible means of conserving the wildlife species by providing as much contiguous natural habitat as possible. The recommendation was to expand the wildlife reserves and give legal protection. The recommendation was implemented creating the under mentioned national parks, forest reserves and sanctuaries afresh, to compensate for the area acquired by the human settlements displacing the occupants of the forest and the water bodies, prior to development.

System	Gross Area (Ha.)	National Park/Forest Reserve/Sanctuary Name	Extent (Ha.)
B	151,000	Maduru Oya National Park	58,850.58
		Thrikonamadu National Park	25,019.00
C	70,000	Wasgamuwa National Park	36,947.99
B & C		Flood Plain National Park	17,350.70
Sub Total	221,000		138,168.27
A	92,600	Somawathiya National Park	37,762.19
		Riverine Nature Reserve	824.15
D	108,000	Victoria, Randenigala, Rantembe Sanctuary	42,087.37
		Minneriya – Giritale Nature Reserve	9,452.86
Total	**421,600**		**228,294.84**

CAPTURE SYSTEM IN GROUNDWATER UNDER PARAMETER AND BOUNDARY UNCERTAINTY

Numerical simulation in groundwater has been commonly used as a decision-making tool for predicting flow and contamination transport and designing pump–and–treat systems. However, the available numerical tools are usually insufficient to design aquifer restoration needs and to meet specified management goals because of the tremendous number of trials that have to be repeatedly simulated.

The combined simulation–optimization technique definitely provides quick decisions of well locations and pumping rates, however, previous investigations have all assumed that aquifer parameters are known with complete certainty, which never exists in any field study. To save project resources and to obtain a high probability of achieving project goals, an optimal design should include the integration of simulation, optimization, and uncertainty components.

Methods for the design of capture and containment systems can be developed using linear simulation management models if the hydrodynamic dispersion is ignored. Since there are always insufficient data on the physical parameters associated with pollutant movement through aquifers, another major objective is to decrease the sensitivity of the system to unforeseen changes in errors of the physical parameters. Valocchi and Eheart described a technique to incorporate parameter uncertainty into coupled linear optimization- groundwater simulation models. In this paper, an extension using a three–dimensional numerical model and developing a generalized linkage between the numerical model and optimization schemes via hydraulic gradient control are reported. A real case study using with the investigations of the uncertainties of hydraulic conductivity in pumping the well layer, infiltration rate, and constant head boundary that affect the optimal design strategies was addressed.

SYSTEM DEVELOPMENT

The system consists of three components, namely,

numerical simulation module, optimization module, and uncertainty module.

GROUNDWATER SIMULATION MODULE

The developed linkage between numerical simulation and optimization was a generalized procedure, which can fit both simulation tools in GMS. Femwater is a three–dimensional finite–element model of density–dependent flow and transport through saturated and unsaturated porous media.

The procedures for constructing a simulation model include the development of a hydrogeologic conceptual model, the development of computational mesh, the determination of boundary and initial conditions, the estimation of model parameters and performing the model run, the conducting of model calibration/verification, and the demonstration of pre-processing and post-processing.

OPTIMIZATION LINKAGE MODULE

The optimization module includes the steps of formulating the management problem, generating the response matrix, inputting the objective functions and constraints, solving the optimization problem, and post-processing of optimal results. This module uses the hydraulic gradient control method, which is based on a unit-response matrix, to perform the linkage between the FEMWATER model and an optimization solver.

The main idea for using hydraulic gradient control is to select wells and pumping or recharge rates so that the contaminant plume does not migrate during cleanup. As long as the plume is stationary, contaminated water can be removed, treated, and disposed or discharged. In the absence of hydraulic controls, the plume will migrate and disperse within the existing and future groundwater flow fields.

Minimizing well rates could mean to extract the most water or to inject the least water. Objectives based on well rates may include all or select managed wells. Once the management problem is converted into a mathematical formulation via response theory, it can be written in a format suitable for input into a linear or nonlinear programming solver. The original code, the revised simplex method, was adopted from numerical recipes and was extensively modified to conduct the construction of this module.

UNCERTAINTY MODULE

We assume that a best estimate of the aquifer parameter, such as hydraulic conductivity, has been in the traditional model described as the management scheme to determine an optimal pumping strategy. However, in some specific regions or layers, the hydraulic conductivity values are uncertain. We used the hydraulic gradient control concept to incorporate the sensitivity criterion into the management model by constraining the maximum possible value of the sensitivity coefficients. The sensitivity coefficients are used to measure the impact of a small change in this parameter upon contaminant plume containment.

MANAGEMENT MODEL WITH UNCERTAINTY CONSTRAINT

The traditional management seeks to minimize the total injection and extraction rate while constraining the hydraulic gradient to be directed inward at the gradient check pair locations. The groundwater flow equation, such as the FEMWATER code, provides the coupling between the decision variable (Q_i) and the hydraulic gradients through the use of a response coefficient. If we assume some specific regions of the aquifer parameter or the boundary values are uncertain, the sensitivity coefficients S_{ik} are used to measure the impact of a small change in this parameter upon contaminant plume containment. The sensitivity coefficient is defined as,

$$S_{ik} = \P P_i / C_k$$
$$\text{where } P_i = h_{Ai} - h_{Bi}$$

is the hydraulic gradient at check pair i, and C_k is the parameter in layer k or boundary value. The absolute value of S_{ik} is a measure of the robustness of any particular management strategy. That is, the strategy with small values of ABS(S_{ik}) will still successfully contain the contaminant plume even if the parameter or boundary value differs slightly from its assumed value. Valocchi and Eheart incorporated the sensitivity criterion into the management model by constraining the maximum possible value of $ABS(S_{ik})$.

The management model is given by,

Nw Objective Function: Min $\Sigma (u_j + v_j)$

- *j=1nw Constraints*: $\Sigma R_{ij} (u_j - v_j) < g_i$ i =1, 2, ... ngp
- j=1 $ABS(S_{ik}) < S_{max}$
- $0 < u_j Q_{max}$ j =1, 2, ... nm
- $0 < v_j Q_{max}$ j =1, 2, ... nm

where uj is the extraction rate at well j, and v_j is the injection rate at well j; g_i is the target hydraulic gradient at location i; R_{ij} is the gradient response coefficient; ngp is the number of gradient check points; nw is the total number of wells; Q_{max} is an upper bound on the maximum possible pumping rate; S_{max} is the absolute allowable sensitivity value. It is noted that the S_{ik} consists of a regional flow component and a gradient response coefficient component.

OPTIMIZATION OF PUMPING TO CONTROL

The developed system with three modules was applied to a restoration site to develop an optimal pumping strategy for determining the contaminant plume.

Problem Identification

This remediation site is approximately 30 acres in size and contains several locations where past spills, disposal practices, and operations have contaminated soil and groundwater. This site and its surrounding vicinity is hydrologically characterized as a valley–fill aquifer system. In response to TCE (trichloroethylene) contamination found in onsite wells, the study site installed a ground treatment facility and is continuing detailed studies of subsurface contamination at the site and its surroundings. Currently, groundwater extraction comes from five pumping wells with four of the five being deep-well pumps, with a daily pumping rate of about 1 million gallons. The treatment system is based on physical separation through air stripping to isolate TCE, with the treated water then being discharged into the river.

Development of a Numerical Flow Model

The numerical model used in this study was based on a hydrogeologic conceptual model. The extent of the esker deposit and geometry of the basement surface are of critical importance to the hydrogeological model. The map module of GMS was used to create a new surface 2D mesh. The basement and sediment geometry with nine different materials classified are shown with the 3D finite element mesh overlay in Fig.3.7 Under the hydrogeologic condition modelled, the mesh is vertically divided into 9 materials and 21 layers. The 3D mesh consists of 11 000 nodes and 20 139 elements.

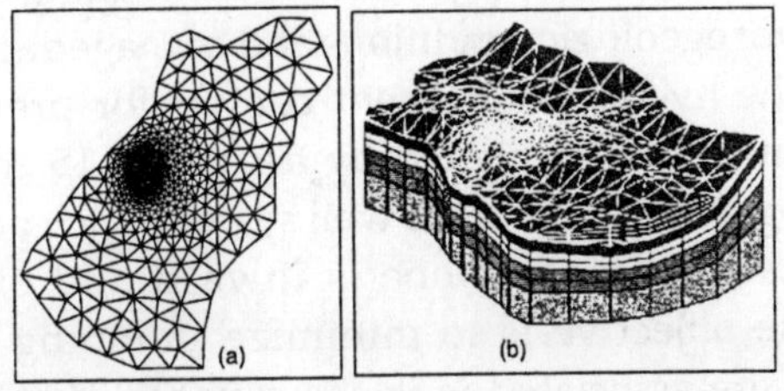

Fig. 3.7 (a) 2D and (b) 3D Computational Meshes.

The western and northwestern model boundaries are represented by the river and are defined by constant fixed head elevations of 116 m. The eastern boundary is a constant head of 165 m. The southern and northeastern boundaries are defined as no–flow boundaries. The recharge and water–well boundary conditions were also assigned to the model. A value of 1.32×10^{-3} m day^{-1} was used as a precipitation influx. Well-pumping boundary conditions were assigned to five locations. A numerical procedure was developed to estimate initial conditions with iteration process due to field measurements were not sufficiently dense enough.

Hydraulic conductivity was assumed to be homogeneous within each model layer. Unsaturated zone curves were required for moisture content, relative conductivity, and water capacity as a function of pressure head. The highest hydraulic conductivity is in the esker layer (86.4 m day^{-1} for horizontal and 50.0 m day^{-1} for vertical).

OPTIMAL SOLUTIONS FOR PUMPING AND MIGRATION

Design of Optimal Pumping Strategies

Since they're presently is no planning for new wells and no injection activities for any pumping wells, hydraulic gradient control is based on the pumping strategies for these five existing wells. With the information from the past 2 years of pumping history, the average demand for each well and the total demand can be estimated.

The plume is located upgradient of the existing well system. Due to the hydrogeologic conditions and management strategies at the site, the hydraulic gradient control for preventing the plume migration was designed by assigning 15 control pairs on the down-gradient side of the well system. Each pair occupies two computational nodes; one is inward and other one is outward. The objective is to minimize pumping rates. Five constraints were associated with the capacity, demand, usage, and individual pumping rates and migration control.

The constraints are as follows:

- Restrict the total volume pumped to allow a low-cost, low–capacity treatment system.
- Restrict the total volume pumped to meet cooling-system demands.
- Restrict the least usage of pumping well 3.
- Restrict individual pumping rates to allow the use of small, inexpensive pumps.
- Restrict heads outside the contaminated zone to values greater than or equal to the heads inside (no spreading).

OPTIMAL SOLUTIONS AND POST

To determine the response matrix, this study only requires

six numerical simulation runs that significantly reduce the computer memory and time required obtaining the optimal solution. The optimal pumping solutions for these five wells are illustrated in Table.

The total pumpage for satisfying the objective function and constraints is 7484 m^3 day^{-1}. This amount is about 86% of full–operation capacity of the entire treatment system. This preliminary result could indicate the current usage must be increased to meet all specified constraints.

Table. Optimal Pumping Rate (m^3 day^{-1}) Compared With Full Capacity and Average Usage for Each Treatment Well.

Well	Average Usage	Full Capacity	Optimal Scheme
1	1171	2286	2014
2	501	1986	1827
3	1566	1566	1050
4	131	1219	1422
5	1261	1589	1171
Total	**4530**	**8646**	**7484**

It is necessary to verify the optimal solutions by checking the optimized values with respect to the desired constraints. Three different variables were used to present this process: velocity field, hydraulic head, and gradient at control pairs. Using the optimal pumping solutions as the pumping rate input and then rerunning the numerical model does this.

The flow field and total head around the pumping and the hydraulic control area show that the resulting depression cone formed around the wells.

Contaminated water will be trapped by these wells and will not be discharged to the river. The clearwater in the upgradient could bypass the contaminated area and discharge into the river. These results show that each control pair under the optimal pumping schemes satisfies the hydraulic gradient control strategy.

Table. Optimal Pumping Solutions (m^3 day^{-1}) the Uncertainty- Constraints of Hydraulic Conductivity, Infiltration Rate, and Constant Head Boundary.

Parameter	S_{avg}	S_{max}	Q1	Q2	Q3	Q4	Q5	Total
No Uncertainty	N/A	N/A	2014	1827	1050	1422	1171	7484
Hydraulic Conductivity	6.2×10^{-3}	6.0×10^{-3}	2095	1913	1089	1282	1218	7698
		3.0×10^{-3}	2228	2264	1151	1565	1293	8296
		1.0×10^{-3}	unfeasible solution					
Infiltration Rate	2.5×10^{-3}	2.5×10^{-3}	2030	1841	1062	1437	1184	7554
		2.0×10^{-3}	2047	1855	1073	1451	1197	7623
		1.0×10^{-3}	2112	1912	1120	1510	1248	7902
Constant Head Boundary	5.5×10^{-2}	5.0×10^{-2}	2132	1973	1128	1539	1241	8013
		2.5×10^{-2}	2242	2165	1174	1673	1283	8637
		1.0×0^{-2}	unfeasible solution					

INCORPORATING THE UNCERTAINTY OF HYDRAULIC

The sensitivity coefficient is defined as the total gradient change when the parameter or boundary value has some uncertainty associated with it. Actually, this total change is based on the regional hydraulic gradient change plus the total drawdown change due to the pumping activities.

Three experiments are investigated, 10% (8.6 m day^{-1}) increase of hydraulic conductivity in the esker layer, 0.01×10^{-3} m^3 day^{-1} of infiltration rate, and 1 m of constant head boundary over the northeaster side of the model increases. These changes are considered as the uncertainty due to the measurement or estimate error. For each experiment, six computer runs are made-one for background run due to parameter or boundary change and five for estimating the response coefficient matrix due to parameter or boundary change. Fifteen estimated sensitivity coefficients for each control pair were used as the sensitivity constraints for solving the optimization problem.

The S_{max} for the tests of each experiment needs to be estimated. Usually, the average response coefficient (S_{avg}) and the average drawdown can be used as the initially estimated S_{max}. The Table summarizes the optimal pumping with given S_{max} for each experiment. Due to the scale difference among these parameter or boundary changes, the optimal pumping scheme changes are difficult to compare. However, for each individual parameter or boundary value, these results are very useful to determine what is allowable error in order to obtain reliable optimal pumping values. For those more sensitive control pairs, it might be the indication for more field measurement to support the reliable study. That some S_{max} obtain unfeasible optimal solution is due to the limitation of capacity constraint in the system.

A very effective and computationally efficient scheme was developed and then applied at a restoration site to determine the optimal pumping alternatives for groundwater remediation. An optimization procedure using GMS and FEMWATER was developed that could be used to contain the contaminant plume while minimizing pumping rates. The sensitivity constraint is used to examine the uncertainty of parameter or boundary value. Considering the capacity, demand, individual pumping limit, usage priority, and migration control of contamination, the results indicate that the optimal total pumping (7484 $m^3 day^{-1}$) was less than the full operation capacity (8646 m^3 day^{-1}) at this demonstration site. However, from the optimal results with uncertainty constraint, it indicates that if the maximum sensitivity of hydraulic conductivity is less than 1.5×10^{-3} or the maximum sensitivity of constant boundary head is less than 1.0×10^{-2}, total optimal pumping might need more than the water with full–operation capacity.

4

Analysis of Forces

INTRODUCTION

The top elevation of a bulkhead project located in Zhe Jiang province was strictly limited in raising at will, in order to match the existing road elevation fixed by the municipal planning. At the same time it's also not the desire to lower the bottom elevation of the parapet for the sake of heightening its weight and stability, because the core of bulkhead was already completed beforehand by means of explosive rock filling and any digging up of it should be avoided as possible. But the extreme high tide level with return period of 50 years which has an important bearing on the safety of the bulkhead are relatively high and the design wave height with accumulated frequency of 1% are also relatively large.

In this condition how to ensure the safety of the parapet becomes a difficult and crucial problem in the design process.The restrain to both upper and lower elevation of parapet slab makes it the only choice to search the space for slab along the horizontal direction rearward. From the view point of usage there will be a sightseeing platform behind the front line of bulkhead within 23m on which overtoppings are permitted, so it's possible to deal with the rear area as part of parapet slab rather than ordinary road work. Under these something special boundary conditions a special form of parapet slab with lower elevation and longer dimension along the direction of wave spreading has put forward out.

As for the uplift forces exerted on the bottom surface of a conventional parapet on the top of bulkhead, it can be calculated and determined using the methods provided by the Code of Hydrology for Sea Harbour, issued by Chinese Ministry of Communication. The calculation methods assume that the distribution of uplift force on bottom surface were in triangular form, *i.e.* the pressure on front toe of slab equals to the average wave pressure acting on the screen wall surface, and that on the rear end of slab approaches to zero.

As calculated results the above uplift force should yet be reduced by 30%.If the elevations of slab bottom keep enough height from the design water level and the length of slab along wave spreading direction are in same order of magnitude with the height of screen wall, the above assumption may be reasonable.But in our case with lower elevation and longer slab of parapet, the current code doesn't meet the need of design. To search the basis for design of new alternatives, the wave model tests were conducted in an irregular wave channel.

Besides the important data such as stability and overtopping discharge etc. have been collected in both regular and irregular wave condition, the pressure distributions both on top and bottom of the new type of slab have also been determined and compared with that of conventional parapet structures.

TYPES OF PARAPET SECTION AND THE BOUNDARY CONDITIONS IN TEST

There are two types of unconventional parapet slab, *i.e.* thick and thin slab and defined as alternative ½and½½ respectively. These two alternatives are same in their dimension and top elevation, the only difference lies in that the slab thickness of alternative ½is 1.5m and alternative ½½ 0.8m. The whole parapet slab can be separated into two parts of front and rear. The bottom surface of the former is equipped with two sills one after another and the seaward section from front sill has a bottom elevation of 2.80m. On the surface of the rear part of slab, there exists a rear wave screen. The two parts of slab are same in top elevation of 4.50m. The front face of parapet is an 1:1 slope, that is consistent with the gradient of armor surface. The armor units are hallow squares with four legs above -1.0m, but are dolos at the foot of slope.

To measure the pressure distribution along the parapet slab, two rows of total 17 transducers were set longitudinally upward (at top plane) and downward (at bottom plane) respectively. Those downward were for the measurement of uplift forces and the pressure directions towards the transducer surface were defined as positive. The transducers at top plane with identical position were for the measurement of forces caused by overtopping discharge and the pressure directions were also defined as positive as towards the transducer surface.

The design water level are extreme high water level of 3.01m with return period of 50 years and design high water level of 2.05m, to which wave height H1% of 4.15m and 3.75m correspond respectively. Pressure measurements have been conducted under regular wave action with wave flatness of 2014 and 11 at each water level to determine the effects of wave flatness on the pressures. These wave flatnesses correspond to wave periods of 11.4s, 8.10s and 6.57s at extreme HWL, but 11.4s, 7.90s and 6.40s at design HWL.

ANALYSIS OF PRESSURE ON THE PARAPET SLAB

The measured pressures for two alternatives at the extreme HWL of 3.01m are listed in Tab.1 taking wave period of 11.4s for example. The non–synchronic pressures in the table indicate the average peak values taken from the pressure–time processes of relevant transducers, so that the diagram plotted by connecting these pressures represent the envelope of average pressure peak values. The synchronic pressures mean that integrating the pressures of all transducers at every synchronic instant in a test run, during which over 2000 sampling data recorded for each transducer, calculating the average and maximum peak values from time process of integrated results, the pressures which are corresponding to maximum peak value are exactly the synchronic pressures showed in the Tab.

The length of slab greater obviously than conventional one, the exerted forces on the slab demonstrate following characters:

- The uplift pressure on the rear end of slab are still not going to reach zero even if the length of slab reaches 1/4 of wave length. The rubble within the core under armor layer keep certain porosity, which allow the pressures produced by wave motion to spread along the wave direction in a long distance. Therefore the uplift pressures on slab bottom near SWL damp out slowly, the stronger the porosity of core is, the slower it damps out. Tab.1 shows that for alternative the non-synchronic uplift pressure on the bottom 20m apart from the front line is 16.12 kpa and is 59% of the uplift pressure on the front end; the synchronic uplift pressure on the rear end may reach 50% of front end.
- The distance between the elevation of slab bottom and SWL has greatly affected on the uplift pressures. At the same extreme HWL of 3.01m, because of the bottom elevation of alternative | | has been increased by 0.17 times of wave height, the non–synchronic uplift pressure at rear end has been 63% reduced compared with alternative in same position. As a result this uplift pressure reaches only 5.94kpa, equals to 25% of the pressure at front end. The synchronic uplift pressure at rare end has also been 57% reduced compared with alternative I.

  Not only the uplift pressure at rear end decreases rapidly by the raise of bottom elevation, but also the total uplift force on the bottom be reduced correspondingly. At extreme HWL the net uplift force on the front part of alternative | | is 21.92t/m, because of the bottom elevation of alternative | |0.17H increased, the net uplift force reduce to 13.52 t/m, which is 38% less than alternative I .As for the net uplift force on rear part of slab, 77% of it has already eliminated.

- The damping rate of uplift pressure is related to wave flatness L/H. Tab.1 is an example in wave period T=11.4s that means L/H=20. As wave flatness is smaller or wave form reveals steeper, the uplift pressure on bottom at rear end would damp slowly. For example of the wave with L/H=11, the pressure at rear end of alternative | is 15.50kpa, which is 72% of the pressure of 21.48kpa at front end; but for alternative the pressure at rear end is 50% of that one at front end. These percentages are all higher than the percentage of 59% and 25% respectively in case of L/H=20. Of cause this is only true as dealing with the ratio of change of uplift pressure. So as to the total uplift forces the one induced by wave with larger wave flatness would be generally greater than the one with smaller wave flatness.
- Because of the lower elevation of slab, the downward exerted forces of the overtopping discharge on the top of structure have played an important role, see the column of "downward pressure on top surface" . At extreme HWL, the ratio of downward force on front part of alternative | | to the synchronic uplift force is 30% in average; but on rear part the ratio reaches more than 0.5. So that if comparing the net uplift force which take the force caused by overtopping discharge into account with those which put the overtopping discharge out of account, the ratio would be 0.84–0.86 for alternative? and 0.60–0.80 for alternativev | |. This shows that the overtopping discharge makes net uplift force cutting down obviously. But in view of the unstability of the effect of overtopping, it should be handled with great care whether take these forces into account in designing.

- In front of parapet structure the wave screen has been canceled and as substitution a rear wave screen be built about 20m behind the front line. In case of lower structure elevation, the wave screen on front line has no distinct function for diminishing the overtopping discharge, on the contrary enhanced the horizontal wave force.The upper elevation and geometric form of alternative I and II are same, at the extreme HWL of 3.01m, the wave pressures on front side in horizontal direction equal to 18-25kpa, which are equivalent to 0.44–0.61γH, and the pressures on rear wave screen are 14–27kpa,equivalent to 0.34–0.66γH. Comparing with the parapet structure with 1m high front wave screen but identical top elevation of slab and in same wave condition, the horizontal pressures with wave screen may reach 44.9kpa, equivalent to 1.1–1.2γH. That means the structures like alternative I or II could not only reduce 40%-50% of horizontal forces in front, but also produce evident phase shift with the forces on rear wave screen.
- The parapet structure with conventional front wave screen would sustain greater uplift pressure than those alternatives in this paper in same condition. As L/H=20 and at two different water levels of 3.01m and 2.05m respectively, the average uplift pressures on bottom of parapet slab with front wave screen are 41.42kpa and 21.86kpa, equivalent to 1.02gH and 0.60gH respectively. But for alternative I the uplift pressures in same condition are 27.21kpa (0.67gH) and 21.66kpa(0.59gH), for alternative II the pressures are even smaller and only reach 25.08kpa (0.62gH) and 12.53kpa(0.34gH) respectively. This is owing to the cancellation of front wave screen, so that the pattern of wave motion in front of structure have been changed from strong reflection and turbulence as well splashing severely to a relatively smooth going overtopping. The reduction of horizontal wave

pressure means the reduction of corresponding uplift pressure, which is identical with the approach of calculation for parapet in general. It should be noticed that if the slab stretches longer and above the SWL with certain distance, the assumptions of uplift pressure distribution being in triangular form and at the rear end of slab the pressure approaching to zero might over estimate the total force, because the uplift pressure might be already decreased to zero before it transmitted to the rear end along the slab bottom.

It should be pointed out that, the problem of parapet structures in this project are extremely large overtopping discharge.

Comparing with the parapet with 1m high front wave screen and in same top elevation of slab, the overtopping discharges induced by irregular wave test are shown in Tab. In the table the row "with 1m wave screen" indicates the measured overtopping discharge over the top of screen near front line, and the row with "alternative I or I I" means the measured overtopping discharge over the top of rear screen about 20m apart from the front line.

- In case of the core materials with certain permeability, for the parapet slab whose elevation is set lower and whose bottom surface lies close to SWL, *i.e.* the distance between them is less than 0.2H (H notes wave height), then on the rear end of slab bottom the uplift pressures would not approach to zero even if the length of slab reaches 1/4 wave length. But if the distance between bottom surface and SWL is larger than 0.4H, the assumption of uplift pressure distribution in triangular form approaches the reality.
- The above distance has great influences to the uplift pressure on slab bottom. The enlargement of the distance leads the ratio of pressures on rear and front end to a rapid reduction, and also to a rapid reduction of the uplift force upon whole bottom plane.
- As the top elevation of slab set lower, there exist two

effects: the advantageous effect of the downward forces induced by overtopping discharge to eliminate the net uplift forces; and the unfavorable effect of enhanced overtopping discharge on the inner region behind wave screen.

- The rear setting of wave screen could evidently diminish the horizontal wave force acting on the whole parapet structure.

ANALYSIS ON DAGUHE GROUNDWATER

The Qingdao City is one of the major cities with serious water shortage in the north part of China. The mean annual precipitation is 680mm and the averaged annual water supply per person is about 370m^3. As a result of continuous economic growth, rapid increase of population and highly improved living standard, the water demand is continuously increased. And water shortage will seriously restrict the economic development.

This requires an effective water resources utilization and water supply management for a sustainable development. Daguhe Groundwater Reservoir is carried out to mitigate the water shortage in Qingdao City.

This project consists of two major parts:

1. Building underground diaphragm wall to cut off the hydraulic relationship between groundwater and sea water;
2. Carrying out artificial recharge projects to replenish groundwater resources.

In addition, the monitoring system, ecological and environmental protection system as well as management system are to be established for the groundwater reservoir. The Daguhe Groundwater Reservoir is located in the plain region of middle and lower reach of Daguhe River Basin. Its length from north to south is 51km and averaged width from west to east is 5–7km. covering 421.7km^2 with an alluvial gravel layer of 5.19m for groundwater storage. The total capacity is 384million m3. The control mode of "4–dry–year–1–wet–year" will be adopted for the groundwater exploitation. This paper analyses the characteristics of Daguhe Groundwater Reservoir and uses two-dimensional mathematical model to simulate the groundwater flow field. To improve the groundwater flow model, a parameter calculation process is also proposed. The results of this model are verified by a 3 years recorded data and applied in the fundamental model of Qingdao water resources macro-management.

GROUNDWATER FLOW MODEL

As part of the fundamental model of Qingdao water resources macro-management, groundwater flow model for Daguhe Groundwater Reservoir simulates the groundwater flow field of the reservoir and provides basic data of water information for analysis and assessment of the Daguhe Groundwater Reservoir.

The model calculation programme are based on the Software of MODFLOW, which is a modularized finite difference solution for groundwater flow model programmed by Michael G Mc. Donald and Arlen W Harbaugh from U.S.G.S.

Conceptual Model of the Water Storage Layer

The water storage layer in this area is a single–layer groundwater system. The groundwater flow can be treated as a two-dimensional non–stabilized underground flow. Due to different structures in different places, the capacity for water storage and water penetration differs in the region. The east and west boundaries are confining boundaries. And the south and north boundaries are open boundaries.

The bottom of the water storage layer is the water-tight stratum. The water storage layer is opened to the air. The replenishment of the groundwater is mainly from the precipitation and influent stream feeding. Discharge of the groundwater is mainly by means of exploitation.According to the above analysis, the groundwater flow model for the Daguhe Groundwater Reservoir can be described by the following formula:

$$\frac{\partial}{\partial x}[K(H-B)]\frac{\partial H}{\partial x}+\frac{\partial}{\partial y}[K(H-B)]\frac{\partial H}{\partial y}+W=\mu\frac{\partial H}{\partial t} \quad x,y\in D,t>0$$

$$H(x,y,t)_{t-0}=H_0(x,y) \quad x,y\in D$$

$$K(H-B)\frac{\partial H}{\partial x}|\Gamma_2=-q(x,y,t) \quad x,y\in\Gamma_2,t\geq 0$$

$$H(x,y,t)|\Gamma_1=H_1(x,y,t) \quad x,y\in\Gamma_1,t\geq 0$$

where:

- H–groundwater level
- B-bottom level of the groundwater storage layer
- μ–storage coefficient of the groundwater storage layer
- K–percolation coefficient
- W– vertical water exchange quantity for the groundwater storage, including Q_{rain}, Q_{stream}, $Q_{evaporation}$ and $Q_{exploitation}$
- $\Gamma_2$second type of boundaries
- $\Gamma_1$first type of boundaries
- District area available of permeation

BOUNDARIES AND GRID DIVISION

The boundary conditions are determined by a statistics of the characteristics of the groundwater storage layer. The east and west boundaries are the natural boundaries of the layer and groundwater storage stops at the boundaries. The north boundary is located at the end of the river valley, which is a feeding boundary, and the south boundary is located at the dividing line of the freshwater and seawater.

The vertical boundary is the bottom of groundwater storage layer, which is composed of clay rock, shale or malmstone. According to the distribution of the water monitoring bores and the slope of the calculation area, the area are divided into grids. The length of the cell border is 1km. There are totally 470 cells in the computation area.

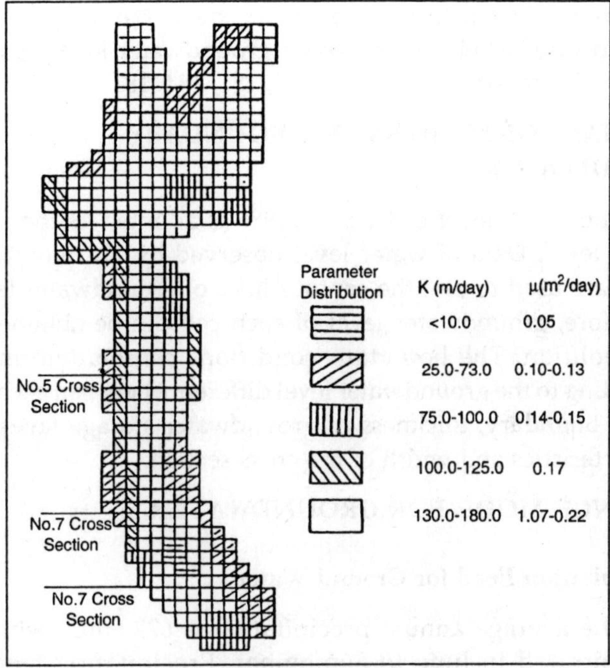

Fig.4.1 Cells Division of Daguhe Groundwater Reservoir and Parameter Distribution

PARAMETERS OF THE GROUNDWATER STORAGE LAYER

The whole area is divided by different sectors according to the different layer parameters, which is determined by the characteristics of the soil and the experimental data in a previous survey by water pumping. According to the survey, the parameters of the groundwater is depicted in Fig. 4.1

ADJUSTMENT OF PARAMETERS ACCORDING TO OBSERVED DATA

To adjust the parameters in the model, the observed data from Dec. 30, 1986 to Dec. 31 1987, which is comparatively more reliable and more complete, are adopted. Series of time interval are 31, 28, 31, 30, 31, 30, 31, 31, 30, 31, 30, 31 days. To verify the parameter values, the observed data from 1988 (low groundwater level) to 1990 (high groundwater level) is used. Series of time interval are 182, 184, 181, 184, 181, 184 days.

INITIAL CONDITIONS AND BOUNDARY CONDITIONS

The water level of Dec. 30, 1986 is adopted as the initial water level. Data of water level observed by 112 monitoring bores are used to plot the contour lines of groundwater levels. Therefore, groundwater level of each cell can be obtained by interpolition. The boundary conditions can be determined according to the groundwater level difference between each side of the boundary, thickness of groundwater storage layer, soil characteristics and width of the cross section.

CHANGE MODE FOR GROUNDWATER

Precipitation Feed for Ground Water

The average annual precipitation is 672.6mm, which is concentrated in June to September. Precipitation feed for groundwater can be calculated by the formula $Q_{rain} = \alpha \times P \times F$, where is the coefficient, P is precipitation and F is area. can be determined by the observed data.

Influent Stream Feed for Groundwater Storage Layer

$Q_{stream} = C_{r\times} K \times \Delta H \times L \times B/N$, where: C_r is the coefficient; K is percolation coefficient for the soil layer under the riverbed; ΔH is the difference between stream water level and the groundwater level; L is the stream length across the cell. N is the thickness of the soil layer under the riverbed. Due to different conditions of the soil of riverbed, parameter value differs in different cells. In this model, three types of parameter values are applied to the cells. The parameter values are calculated out with the data of three cross sections of the river.

Extracted Groundwater Quantity

The average water extraction for the whole region, which is

obtained from the total water extraction and cell numbers, is used as the calculation groundwater extraction for each cell,

Groundwater Evaporation

Observed data for surface water evaporation in Nancun Station is used as the maximum groundwater evaporation. For each cell, the following formula is used to calculate the groundwater evaporation:

$$Ret = \begin{cases} R_{etm}, & \text{when } h>h_s \\ 0, & \text{when } h<h_s-d \\ R_{etm}\,(h-h_s+d)/d, & \text{when } (h_s-d)<h<h_s \end{cases}$$

R_{etm}–the maximum groundwater evaporation
h–groundwater level
h_s–land level
d–the maximum depth for groundwater evaporation

For each cell, h_s can be obtained by the topographical map and d is 3.0m according to the local soil condition.

CALCULATION ON PARAMETER OF STREAM FEED FOR GROUNDWATER

To improve the model for groundwater flow, parameter of the influent stream feed for groundwater should be calculated and verified. Due to different conditions of the soil under riverbed, parameter value differs in different cells.

In this model, observed data for three cross sections of the river are used for the parameter calculation, which is an improvement to the estimated parameter of Qstream. As described, Influent stream feed for groundwater can be calculated as the following:

Qstream =Cr′K′DH′L′B/N, which can be simplified as: Qstream =Cri_v′DH, where Criv= Cr′K ′L′B/N, DH=Hr–Hgw, where Hr is the stream water level and Hgw is the groundwater level. In case of a too low groundwater level, permeation stops above the groundwater level where Qstream can not be described as Qstream =Criv′DH. In this case the maximum groundwater level is Hbot.

Therefore Qstream can be described as the following:

Qstream = Criv ′(Hr–Hgw) when Hgw > Hbot Criv ′(Hr–Hbot) when Hgw < Hbot where Criv can be defined as the parameter of stream feed capacity for groundwater. To calculate the parameter value of Criv, an observed data series of DH and Qstream are provided. The Criv values for the No.5 cross section, No. 7 cross section and No.8 cross section are worked out by linear regression of DH and Qstream. Fig. 4.2 shows the calculation case of No.7 cross section. The results of the parameter calculation are showed in part 4 of this paper.

Results of parameter calculation:

Criv5=20575.9m^2/d, C_{riv}7=4331.5m^2/d, C_{riv}8=16000m2/d.

When applying the values into the whole model of Qingdao water resources macro-management, the result of total stream feed for groundwater is 74.49million the year of 1985, which is in accordance with the observed data of 71million. By applying the parameter into the groundwater flow model for Daguhe Groundwater Reservoir, groundwater level can be obtained at any given time and can be used for predicting the groundwater level.

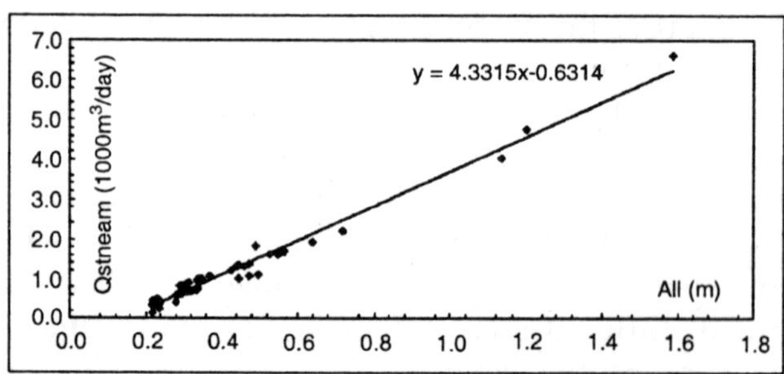

Fig. 4.2 Linear Relationship between Q-ΔH For No.7 Cross Section

To simulate the groundwater flow in Daguhe Groundwater Reservoir, a two–dimensional model is applied. Finite difference method is applied for the solution of the model and parameters are determined and adjusted according to observed groundwater flow data. To improve the model, parameter of the influent stream feed for groundwater are calculated out. Observed data for three cross sections of the river are used for the parameter calculation, which is an improvement to the estimated parameter value.

The results of this model are tested by the 3 years observed data and the model are successfully used in the fundamental model of Qingdao water resources macro-management.

APPLICATION OF THE SURFACE WATER AND THE GROUNDWATER

The surface-water and Groundwater co-regulation developed rapidly with the development of the modern industry agriculture and scientific technology because the global population is increasing so that the demand for the water quantity is growing, which cause the contradiction of the supply and the demand for water source. The surface-water and the groundwater co-regulation makes the utility of the whole water resource system optical by controlling the hydraulic factors of the surface-water and groundwater.

Haimes et al dissociate the surface water and the groundwater, and study the optimal management of distribution in the water use zones of surface water and the extracting in the water use zones in groundwater by using the dynamic and linear programming.H.J morel–seytowy and Tissa Illangsekave built the functional relation of the control decision variable, initial state and result state of river–groundwater by series of response function. At the same time, they studied the model of the river–groundwater co-regulation. In homeland, Xu Juanming et al first built the model of the surface water and the groundwater co-regulation in Qing Huangdao Shi River and manage. In this model, the groundwater is regarded as the concentrated parameter.

FEATURE OF THE SURFACE WATER SUBSYSTEM

Shifosi reservoir is a sole important control engineer in Main River of Liao river.Its task is the flood prevention and the water supply for Shenyang.It will be as an equalizing reservoir in the future. This reservoir is a plain reservoir, which retains water by river weir, big dike in the north of Shenyang, flood protection embankment of the both Liao riverbanks, tributary backwater dike and surrounding village dike. Its normal pool level is 47.0 meter and the maximum flooded area is 44.0 km^2. Its design reservoir capacity is 1.187×10^8 m^3.

The catchment area from Shifosi reservoir to the source is 1.655×10^5 km^2, which is converged by east and west Liao River and their tributary. The mean annual net inflow volume is 11.937×10^8 m^3.

It has two features:

1. Annual distribution is not very average and main inflow is in July, October and September.The net inflow volume is 11.937×10^8 m^3 which is per 80 of the annual inflow volume.
2. The overyear variation is obvious. The most net inflow is 40.711×10^8 m^3 in 1964 and the least is only 0.357×10^8 m^3 in 1982.The former is 114 times as much as the latter. So if Shifosi reservoir is used as the only water supply resource, the utilization rate is very low.

Shifosi reservoir is a plain reservoir which is built on the Quaternary system's friable sedimentary layer. The elevations of the main dam and the big dike are 51.09 m and the most reservoir water level is 47.0 m. The river bed height mark of the dead water level at the main dam base is 38.3m.The river safety flood discharge Lower River of Shifosi is 5500 m^3/s.

After the reservoir will be built, the flood control standard, which was once in 20 years now, is increased to once in 100 years. The peak flow is 8991m^3/s. As for the reservoir, the water is ungated from 1 st July to tenth September annual when there is no water surplus except overflow in river course. So the reservoir can't supply water.

FEATURE OF GROUNDWATER SUBSYSTEM

In this region, the groundwater is the pore phreatic water or the pore micro-confined water in the Quaternary system's friable sedimentary which is shallow. The water-bearing media is main the run-flood sand and the gravel in medium–upper Pleistocene series. In longitudinal direction, the particle size turns coarser and coarser from upper the bottom. The top is medium-fine sand.

The middle and the low are coarse sand and sandy gravel whose general thickness is from 17 m to 30 m. Near the Liao River, the thickness grows thicker and the most thickness is 42 m. From Liao River to its bilateral region.the thickness turns thin. In transverse direction, from Liao River to east west both sides the particle size of the water–bearing media gets more and more fine.

The top of the most part of the water-bearing system is a layer of the subsand soil and the loam whose thickness is general from 1 to 10 m. Near Liao River, the layer gets thin or is deficient.So the groundwaters of the region is phreatic or shallow confined. The permeability coefficient of this aquifer is from 20 to 90 m/d. The average is 53.4 m/d. The unit water boil volume is from 600 to 2400 m^3/d.m.

The input types of the groundwater system are natural input and artificial input. The former includes the precipitation infiltration, evaporation, influent seepage recharge and discharge of the river channel, inflow and outflow of the lateral run-off. The latter includes extracting groundwater for industry agriculture and the infiltration of the irrigation water.

The groundwater levels as a kind of output form exits two peak values every year. The first peak value appears in August and September when the rainfall infiltration is strong. The second peak value which is less than the first one appears in February and May when the snow water infiltrates to recharge the groundwater. The hysteresis period of the groundwater infiltration responding is general from several days to a month.

The minimum level appears in from May to July every year when the vertical infiltration is very small because under the artificial extracting groundwater and the evaluation discharge, the groundwater is in the dissipation. The annual range of the ground is general from 0.5 to 0.3 m and the annual range of the central extracting region is from 8 to 9 m.

THE CO-REGULATION MODEL

In the light of the total demand for the design planning on Shufosi reservoir, we should consider three rules as followed:

1. Make full use of surface-water; discrease possibly the reservoir surplus.
2. Utilize the interference wells and the regulation function of aquifer space and sure the aquifer not to be unwatered locally.
3. Under the premise of not causing to submerge the reservoir, first utilize the surface-water (the embedded depth of the water level doesn't reach 1 meter as the immersion standard)

Building of Co-Regulation Water Volume Model

Shifosi reservoir keeps the close hydraulic relation to the groundwater in aquifer out of the reservoir by the permeable layer under dam (or dike) and the low water-bearing of the reservoir bottom. The reservoir leakage supplies the aquifer. The variation of the reservoir level effects the leakage. At the same time, the extracting volume of the interference wells out of the dike determines the variation of the groundwater in aquifer out of the reservoir. They effect each other. Under the regular structure of the groundwater system, the variation of the reservoir level and the groundwater determines the exchange quantity of the surface water reservoir and the groundwater reservoir by which the surface-water and the groundwater level are united a system.

Solution of the Model

In the joint water quantity model, the surface water dynamic model is the linear equation in which the net inflow is known and the other is unknown. But the latter is relate to the reservoir capacity of each time interval which is the function of the reservoir water level which is solved. The groundwater model is a large non-linear equation group and can't solve directly. So for the groundwater model, we use the over relaxation iteration method and for the surface water, we use Seder iteration method.

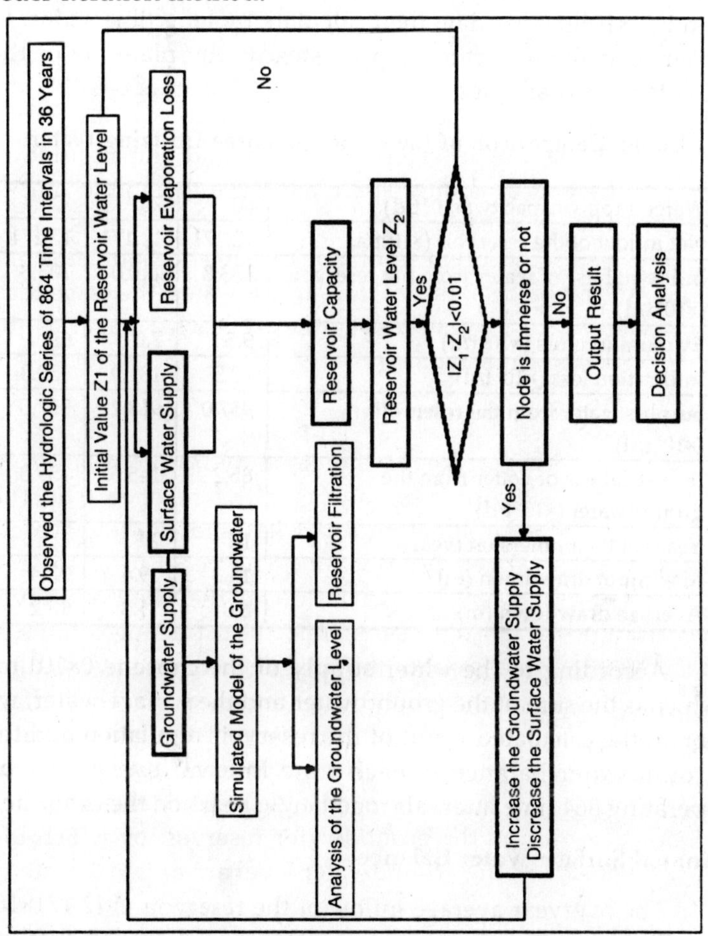

Fig. 4.3 Chart of the Calculated Programming of the Groundwater and Surface Water

ALCULATED RESULT ANALYSIS

Calculated result analysis of the chosen scheme

According to the reservoir net inflow, time distribution,

aquifer structure, and extracting condition of the groundwater, the chosen water supply are respecteve 70×10^4 m^3/d, $72\times10^4 m^3/$ dand 60×10^4 m^3 per day, then calculate by long time series. In order to make the water supply steady, the planning of the 70×10^4 m^3/d is adopted.

Table. Comparison of the Result of Three Utilizing Water

Water supply capacity ($\times10^5$ t/d)	60	70	72
Net inflow of the reservoir ($\times10^8$t/a)	12.171	12.171	12.171
Industrial use of water from the reservoir ($\times10^5$ t/d)	133.8	110.0	103.3
Evaporation loss ($\times10^5$ t/d)	9.5	7.0	6.2
Infiltration loss ($\times10^5$ t/d)	70.3	131.9	150.3
Surplus water from the reservoir ($\times10^5$ t/d)	957.0	968.2	957.0
Industrial use of water from the groundwater ($\times10^5$ t/d)	85.2	145.5	159.3
Years of the immersion (year)	16	2	1
Maximum drawdown (m)	14.3	19.4	22.4
Average drawdown (m)	6.0	11.3	13.2

According to the water supply of the chosen 70×10^4 m^3 which is the sum of the groundwater and the surface-water, we obtain the calculated result of the reservoir regulation and the groundwater balance in each time interval every year by operating 864 time intervals together 36 years on the computer.

Annual Surface Water Balance

The overyear average inflow of the reservoir is 12.171×10^8 m^3. The overyear average surface–water supply is 1.100×10^8 m^3 (30×10^4 m^3/d) which are per 43 of the total industrial use of weter.

The overyear average groundwater supply is 1.461×10^8 m^3 (40×10^4 m^3/d) which occupies per 57 of the total industrial use of water, The reservoir evaporation loss is 0.07×10^8 m^3 and the infiltration loss is 1.319×10^8 m^3 of which 0.202×10^8 m^3 is the dam base infiltration loss, of which $1.17\times10^8 m^3$ loss supplies the groundwater. The reservoir surplus water is $9.682\times10^{8}m^{3}$ which is per 79.55 of the total inflows. Per 77 of the extracting of the groundwater comes from the recharge of the reservoir.

Annual Groundwater Balance

The overyear average lateral recharge is 0.164×10^{8} m^{3}. The vertical recharge is 1.223×10^{8} $m^{3}$and the reservoir water recharge is 1.319×10^{8} m^{3}.

So the total recharge is 2.706 $x10^{8}$ m^{3}. The lateral discharge is 0.041×10^{8} m^{3}. The vertical discharge is 0.35×10^{8} m^{3}. The river discharge is 0.0367×10^{8} m^{3}. The origin extracting is 0.598×10^{8} m^{3}. The new increment of extracting is 1.461×10^{8} m^{3}. The total discharge is 2.481×10^{8} m^{3}. So the overyear average balance is $0.219\times10^{8}m^{3}$.

For the balance of the groundwater, we choose three typical years to analyse. In high flow year, the total recharge is 3.334×10^{8} m^{3} of which the surface-water recharge is 1.662×10^{8} m^{3} which is per 49.9 of the total recharge.

The total loss is 2.007×10^{8} m^{3}.The former is more than the latter and the positive balance is 1.32×10^{8} m^{3} which makes the water level lift. In normal year, the total recharge is 3.191×10^{8} m^{3} of which the surface-water recharge is 1.799×10^{8} m^{3} which occupies per 55.7 of the total recharge. The total loss is 2.647×10^{8} m^{3}. The both balance basically.

The positive balance is 0.546×10^{8} m^{3}, which cause water level lift slightly. The groundwater appears the positive balance. In low year, the total recharge is 1.302×10^{8} $m^{3}$of which surface-water recharge is 0.271×10^{8} m^{3}.The total loss is 3.084×10^{8} m^{3}. So the former is less than the latter, which makes the groundwater system appears negative balance whose value is 1.782×10^{8} m^{3}. So the groundwater level drops.

Result Analysis

Under the daily water supply 75×10^{4} m^{3}, the average surface-water is 30×10^{4} m^{3}/d and the groundwater is 40×10^{4} m^{3}, we may obtain the rules of the surface-water level and groundwater level of 864 time intervals in 36 years. The maximum value of the groundwater level appears in 2022 which is a high flow year.

Except that the drawdown of the groundwater level, which is the upper reaches off the reservoir bottom, is no more than 1.0 m, other is more than 1.0 m.

So it will not appear immersion of near area of the reservoir by operating this planning. The average drawdown depth of the groundwater level is 8.4 m which is third-one of the average thickness of aquifer. So the surplus depth is 15 m, which will not effect the operation of water pumps, what is more, the aquifer is not unwatered which meets the general co–regulation principle.

The general variable rule of water level: Because in low year, the part storage of the groundwater system is used, the water level falls. The groundwater obtains recharge in high flow year of season, so the groundwater level rises again. From the overyear variation of water level, the groundwater system is basic in dynamic equilibrium.

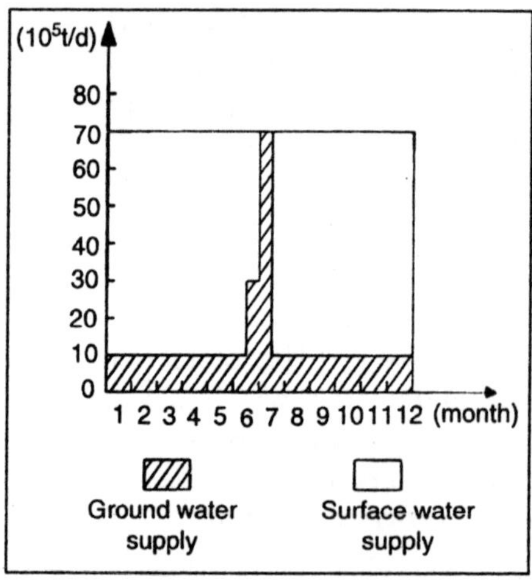

Fig. 4.4 Joint Water Supply Scale Surface-Water and Groundwater

So we see, the planning operation of 70×10^4 m3 will not destroys the ground Water equilibrium and cause the region water level to fall,also not appears immersion, which brings into full play the storage action.

The distribution water supply scale the three high, normal, and low years as followed. The surface water and groundwater co-regulation can supply water of 70×10^4 m^3/d for ShenYang which is reasonable under the several calculation plannings. The co-regulation will not cause the around large mass immerse. The overyear average groundwater is 40×10^4 m^3/d which is the extra extracting on base of the present industrial and agricultural extracting.

The most part of the extra extracting comes from the loss of the reservoir of the river, which raise the utilization coefficient of the surface water. The reservoir infiltration in the normal year is 1.779×10^8 m^3 which is per 80 of the extra pumping. So the co-regulation of the surface water and the ground water will not effect the present water supply.In a word, this example proves the availability of the surface water and the ground water co–regulation.

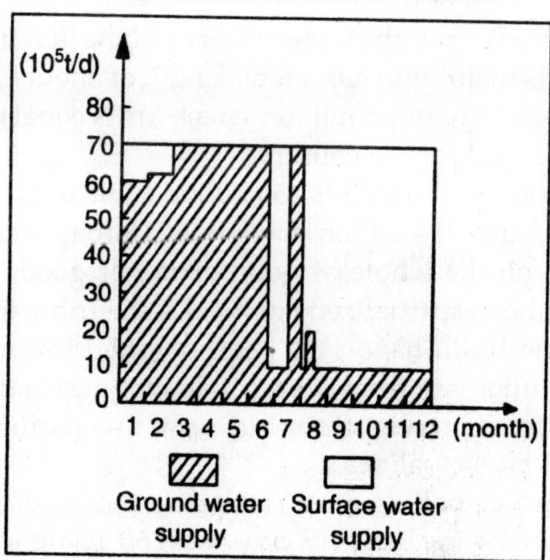

Fig. 4.5 Joint Water Supply Scale of Surface Water and Groundwater in 2007

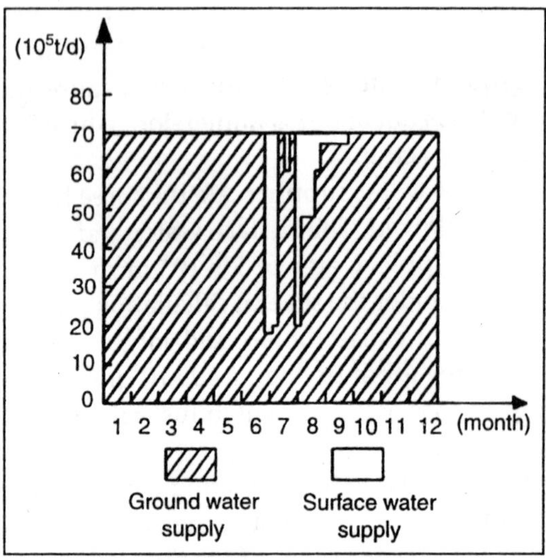

Fig. 4.6 Joint Water Supply Scale of Surface-Water and Groundwater in 2018

BALANCING WATER AND SOCIAL CHALLENGES

The Yangtze River is the largest river in China. It ranks third in the world. Its main stem has a total length of about 6,300km. Its drainage area (D.A) of 1.8 million square km is equal to about one fifth of the total for the country.

The mean annual run-off is 450 billion cubic m at Yichang gaging station, and 914 billion cubic m at Datong station. Its water quality, on the whole, is good or rather good, but the reaches along the industrialized city banks, the tributaries and the lakes of the basin have now been polluted heavily. The sources of pollution are from urban, industry, agriculture and moving sources of navigation, but now are mainly from industries and municipalities.

The pattern of pollution in Yangtze is the occurrence of polluted belts along the banks of cities with 60-100 m in width. The polluted belts are all located downstream from the outlets of sources of pollution. Now the polluted belts are discontinuous in the longitudinal direction of the river. If further pollution should not be checked, the discontinuous polluted belts near the cities would connect into a continuous one.

REASONS FOR POLLUTION OF YANGTZE

The pollution of water in Yangtze basin has not been checked is due to:

- The population expanded dramatically. During the development of urbanization the environmental protection has not been laid as the focus of economic decision making, so that the water and social challenges could not be balanced.
- For management, much of industrial wastewater and the polluted discharge from cities did not meet the standard of emission. The actual river basin management didn't fulfill.
- The regulations and laws were not completed yet. The funds invested were not enough.
- Actual public environmental awareness and self–consciousness on the risks of water pollution, soil erosion and the ecological damage are not so high.

STRATEGY

Adopting sustainable development strategy, which is the harmonious and integrated coordination of economy, society as well as population, resources and environments.

The frame diagram for processing the formulation of sustainable development strategy for industry, agriculture and city, as well as for a region or a river basin, a country, the global area as shown in the left part of the diagram.

The right part of the diagram addresses the environmental protection strategy of the corresponding parts. The whole diagram shows the action and reaction of sustainable development to/from environmental protection in order to make them as an integrated coordination.

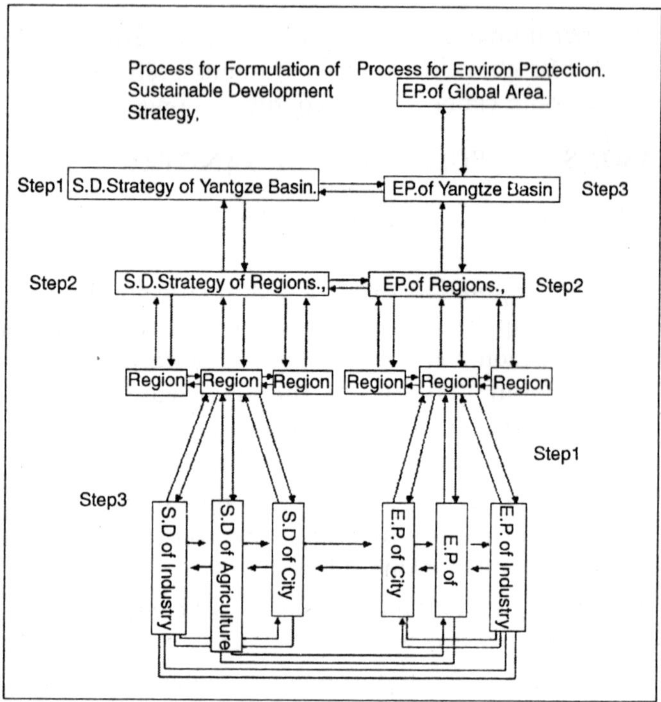

Fig. 4.7 Frame Diagram Formulation of S.D Strategy and Its Environmental Protection

Sustainable Development May be Defined as Three Main Aspects

1. From temporal point of view: as one that "meets the needs of the present without compromising the ability of future generations to meet their own needs."
2. From spatial point of view: as one that meets the needs of specific region could not damage and weaken the ability of related regions to meet their own needs.
3. From the relationship of human being with nature, it requires the two should be survived in harmonic manner.

METHODOLOGY

For planning new industries: by sustainable development strategy:

Case Study 1-integrating the Development Strategies of Riverside Cities with Protection of Yangtze

Convert the superiority of the water resources of Yangtze into the superiority of economic development of the riverside cities is the major trend of history. Abundant water resources and power supply as well as the facilities for navigation are the basic prerequisites for city economic development. Yangtze is supreme in all aspects, so it has many facilities for the development of riverside cities. Ordinary the development will give some adverse effects on environment and on Yangtze. Only by integrating the strategies of development of riverside cities and protection of Yangtze in a harmonic manner, it could break the link between economic growth and environmental damage.

The principles for development and water pollution control must adhere to the following as close as possible, namely:

- Beneficial to the parts concerned, advantageous to the whole area and at least through mitigation, harmless to the other;
- Advantageous to the present, beneficial to the coming period and harmless in the future;
- Helpful in improving the living standards of human being and harmless to the equilibrium of ecosystems of the basin;
- Select the quality objectives of the water environments in accordance with the designed functions of the water body and raise its standard(level) step by step with full consideration of our state condition and financial possibility;
- Control the most polluted reaches, tributaries and sources in priority;

For fulfilling the above principles we have laid the emphasis on the following work:

- *Applying comprehensive pollution control planning and the complex remedial biologic and engineering measure:* The main principle for river pollution control should be to restrict waste effluent to reduce the loading of pollutants and to exploit clear water source to increase the self-purifying(assimilative) capacity of the aquatic environment. In planning to rationally deal with the above two aspect and to coordinate regulation of the river reaches, the whole main stem of the river and the whole basin are very important.

- *Control strictly the pollution sources coming from factories, mines, enterprises and village industries by:*
  - Economic use of water,
  - Diminishing the waste water discharge,
  - Controlling the emission of pollutants by quantity,
  - Good planning and siting for the development of village industries.
- *Take care of the main sources of pollution, and control polluted water from cities by the following measures:* Improving the drainage and sewer system and associated facilities, construction of concentrated sewage treatment plants by municipalities to treat mainly for domestic sewage or construction of a plant by a big factory to treat industrial polluted water and sewage, or construction of a plant by several factories jointly, choose suitable control techniques for each special locality, turn the waste in the polluted water into resources to be reused.
- Rationally site new industries and urban developments, preventing further pollution from new sources and regulating existing sources.
- Extend both industrial discharges and sewage outfalls into the middle of the river in certain reaches.
- Prevent soil erosion to reduce the turbidity and the organic content of the river.
- *Control oil pollution and other pollutants from navigation:* According to Chinese Regulations, the outlets of industrial polluted effluent and domestic sewage discharges should not be located at a distance nearer

than 1000m upstream and 100m downstream from the site of the inlet to a water supply. Ordinarily many alternatives for pollution control of the Yangtze along the shore are formulated among different distances (less than 1000m) downstream from polluted water outlets in order to restore the quality of the river, such as the 300m recovery plan, the 500m recovery plan, the 800m plan and 1000m recovery plan, etc. The shorter the recovery distance, the more expensive is the cost of the plan. The plans selected are different at different times. In general, at present the longer distance recovery plan is adopted, whereas in the future a smaller distance recovery plan will be selected. Of course, the fundamental principle would be to minimizing harmful fluxes from land to water in Yangtze finally.

- *Main environmental problem of water of the existing riverside cities:* Most industries locate at the suburban of the riverside cities along the Yangtze River. Some of the existing riverside cities have different kinds of environmental problems. Their environmental problems of water are also different among cities, but most of them have the problems of city flood control, pollution control of water environment, insufficient city water supply during summer time and insufficient capacity of drainage system for intensive storm water.
- *In summary*: Yangtze is the big support for the development of riverside cities, but it must be reminded that if we should pay attention only to economic development and disregard the protection of Yangtze, the big support would vanish at last and then the development would not go any further. So integrating the strategies for economic development and protection of Yangtze as well as the environment is urgent. To make development of riverside cities keeping pace with protection of Yangtze if necessary, and no doubt it could be obtained. The ultimate status of

environment quality and the state of ecosystems of the cities and the basin are the indices for examining if the economic development strategies are correct or not.

Case Study 2-protecting the Water Quality in the Three Gorges Reservoir

- The Three Gorges Project (TGP) is a key project in the development and harnessing of the Yangtze River. The dam site is situated in Sandouping of Yichang city, Hubei Province, with a distance of 40km upstream from Gezhouba dam. The TGP is a multipurpose hydro-project mainly for flood control, power generation, navigation and water supply improvement. *The main characteristic of Three Gorges reservoir is:*
  - The relative volume of the reservoir is small and its regulation capacity for run-off is low. It's a seasonal regulation reservoir.
  - The reservoir is of gorge type, with a length of 660km and an average width of 1.1km.
  - The operation programme for reservoir regulation considers various requirements including the environmental aspect.
- Now the total wastewater and sewage discharged into Reservoir reach amounts to more than 1 billion ton, the living solid waste of non-agricultural people in Chongqing city amounts to 1.9 million ton, the industrial solid waste amounts to 8.3 million ton annually. They are the main sources of pollution to the future reservoir. Now the water quality of Yangtze, however, remains good in general due to huge quantity of stream flow, except for pollution belts along the banks near cities. After impoundment the slower flow velocity and higher water level cause by the TGP will aggravate shoreline pollution. Therefore, it would be necessary to strictly control the discharge of wastewater from surrounding factories, mines,

towns and cities, to enforce treating the solid wastes and to be mitigated further by better wastewater and solid wastes treatment measure.

Moreover, the impact of inundation on resettlement and the environmental planning for the reservoir have been compiled. In order to protect the environment and to keep the water in the reservoir to be clear, the industries and third estates newly established should be carefully selected. Only those of no or less pollution would be allowed to build in the region.

For the existing industries: ask them to meet the standard of emission in a definite time by the new technology, such as to use the new environmental-friendly inhibitors development circulating cooling weater, etc. For the serious pollution factories: by closing, stopping their production, joining many factories together, turning them to other production or moving to other places, etc. to meet the necessary standard of total pollution loading control of some important locations.

RESULTS AND CONCLUSIONS

The sustainable development planning in China is to treat the river basin as an ecosystem, as an integrated unit and to check the industrial pollution as its main aspect. First to arrange the plan of economic and social development and to investigate both water demand and pollution loading. For the existing industries, asking them to meet the standard of emission in a definite time by the new technology or by some severe control methods according to their conditions. Both are useful and efficient for balancing water and social challenges and could break the link between economic growth and environmental damage

BASE-FLOW ANALYSIS ON HYDRAULIC CONDUCTIVITY USING NUMERICAL

For the efficient use of water resources, it is important to understand the relationships between basin yields and geographical characteristics of a watershed such as basin area, slope gradient, and the permeability of soil and geology layers. One of practical way to understand them is physical consideration on the relationships using distributed run-off model, because they enable us to analyse the effect of a single geographical characteristic on discharges.

Similar attempts were made by many researchers, but they have not made on the effect of anisotropic hydraulic conductivity of soil and geology layers on base-flow behaviour. Hence, this study tried to investigates the relationships between those two using numerical simulations. The model had vertical, two-dimensional, and rectangular cross section and was sloped from the ground surface. The anisotropic soil and geology layers were expressed by employing the hydraulic conductivity in the direction of hill-slope and its normal direction.

NUMERICAL MODEL

Model Structure

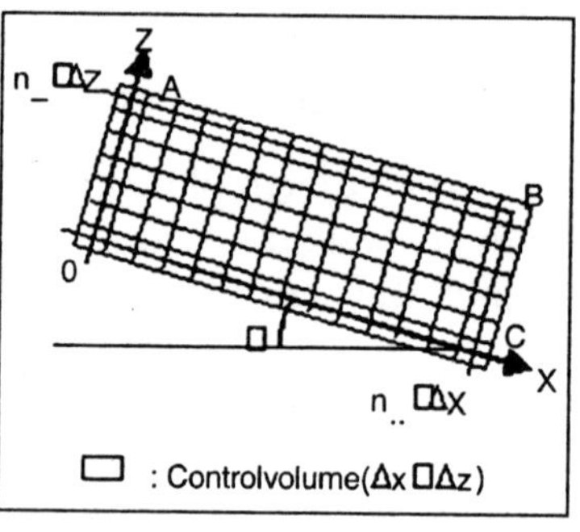

Fig. 4.8 Model Structure

Figure 4.8 shows the structure of the numerical model designed to have vertical, two-dimensional, and rectangular cross section. The model is inclined to α radian from horizontal plane. The segments OA and BC model the upper and the lower end of a watershed, respectively. The segment AB corresponds to the ground surface that has precipitation and evapotranspiration. The segment OC models bottom of the soil and geology layers. The rectangle OABC consists of small cells with the area of Δx times Δz. The x-axis and z–axis are set as shown in Figure 4.8 and nx and nz are the number of cells in the x–direction and z-direction, respectively. Because this model doesn't consider surface run-off, precipitation and storage water flow through the model to the segment BC that yields the discharge from this model.

Governing Equation

The rectangle OABC models soil and geology layers that play important roles for hydrogical processes, and the appropriate governing equation for this model is the Richards' equation as shown in equation.

$$\left(\frac{\partial\theta}{\partial t}=\right)C\frac{\partial\psi}{\partial t}=\frac{\partial}{\partial_x}\left(K_x(\Psi)\frac{\partial\Psi}{\partial_x}-K_x(\Psi)\sin\alpha\right)+\frac{\partial}{\partial_z}\left(K_x(\Psi)\frac{\partial\Psi}{\partial_z}+K_z(\Psi)\cos\alpha\right)$$

$$\begin{bmatrix}\theta:\text{Moisture content},\psi:\text{Pressure head},\alpha:\text{Gradient of slope},x,z:\text{Direction in Fig.1,}\\K_x,K_z:\text{permeability in direction of x and z},t:\text{Time},C:(=\partial\theta/\partial\psi)\text{Specific water capacity}\end{bmatrix}$$

This equation cannot be solved without the relationships between moisture content θ, pressure head Ψ and hydraulic conductivity K_x (and K_z). Equation, Tani's equation, was employed to relate θ to Ψ. The relationship between θ and K_x (or K_z) was modeled by equations.

$$\theta=(\theta_s-\theta_r)\left(\frac{\Psi'}{\Psi_0}+1\right)\exp\left(-\frac{\Psi'}{\Psi_0}\right)+\theta_r,\quad\Psi'=\begin{cases}\Psi & (\Psi<0)\\0 & (\Psi\geq 0)\end{cases}$$

$$K_x=Ks_x\left(\frac{\theta-\theta_r}{\theta_s-\theta_r}\right)^{\beta}\qquad K_z=Ks_z\left(\frac{\theta-\theta_r}{\theta_s-\theta_r}\right)^{\beta}$$

$$\begin{bmatrix}ks_x,Ks_z:\text{Saturated hydraulic conductivity},\beta:\text{Parameter}\\\theta_s:\text{Saturated moisture},\theta_r:\text{Residual moisture},\psi_0:\text{Value of }\Psi\text{ that maximize }C(\Psi')\end{bmatrix}$$

Boundary Conditions

According to the model structure, the following equations were employed as the boundary conditions for the segments OA, CO, BC, and AB, respectively.

$$\left.\begin{aligned}&\frac{\partial\Psi}{\partial x}=\sin\alpha\qquad\Psi=0\qquad(\text{if saturated})\\&\frac{\partial\Psi}{\partial z}=-\cos\alpha,\ \frac{\partial^2\Psi}{\partial x^2}=0\quad(\text{if saturated})\end{aligned}\right\}$$

$$\left[r(t): \begin{cases} \text{Evapor tan spirtation}(mm/h) & (r(t) > 0) \\ \text{Precipitation}(mm/h) & (r(t) \leq 0) \end{cases} \right]$$

NUMERICAL EXPERIMENT

Run-off simulations were performed only changing the hydraulic conductivities in x–direction and z–direction. The other parameters of the model were evaluated as shown in Table.

The pressure head at every point in the model were set at –0.9 m that is equal to the moisture content of 0.4, as the initial condition of the simulation. Precipitation and evapotranspiration were not provided to the model, and only run-off quantities were calculated for 500 days long.

Table. Model Coefficients

Name of parameter	n_x	n_z	Δx (m)	Δz (m)	a (radian)	θ_s	θ_r	ψ_0 (m)	β
Value of parameter	99	9	1.0	0.5	$\pi/10$	0.7	0.3	-0.3	3.5

The simulations can be divided into three groups. The saturated hydraulic conductivity in the x-direction, Ks_x, was only changed while that in the z-direction, Ks_z, was kept as constant in the first group.

The variable in the second group simulations was only Ks_z. Effects of both Ks_x and Ks_z on base-flow were investigated by the simulations in these two groups. The simulations of the third group were carried out to observe the combined effects of randomly evaluated hydraulic conductivities in the two directions on the base-flow.

Result of Experiment

Five cases of experiments were performed by changing the value of Ks_x from 0.002 cm/s to 0.01 cm/s, where the value of Ks_z was fixed at 0.01 cm/s as shown in Table. The result of experiments is shown in Figure 4.9.

Table. Variation of Ks_x

Case	Ksx-1	Ksx-2	Ksx-3	Ksx-4	Ksx-5
Ksx(cm/s)	0.002	0.004	0.006	0.008	0.010
Ksz(cm/s)	0.010	0.010	0.010	0.010	0.010

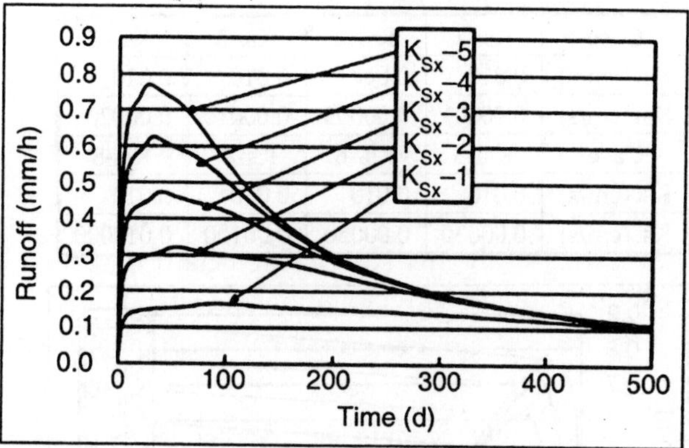

Fig. 4.9 Simulation Result (Ks_x)

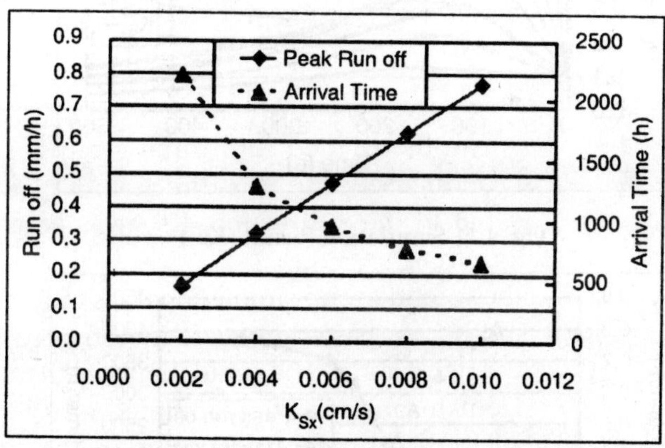

Fig. 4.10 Ks_x, Peak Run-off, and Its Arrival Time

The peak run-off became greater and the recession curves become steep when Ks_x was great, as shown in Figure 4.10. Figure 4.11 shows the relationships between Ks_x and peak run-off and the arrival times. As you see, peak run-off was proportional to Ks_x and the arrival times decreased exponentially as Ks_x become greater. As shown in Table, eight cases of experiments were carried out. Ks_z was varied from 0.00001 cm/s to 0.01 cm/s, while Ks_x was fixed at 0.01 cm/s. Figure 4.11 shows the results.

Table. Ks_z Variation

Case	Ks_z-1	Ks_z-2	Ks_z-3	Ks_z-4
Ks_x (cm/s)	0.010	0.010	0.010	0.010
Ks_z (cm/s)	0.00001	0.00005	0.00010	0.00020
Case	**Ks_z-5**	**Ks_z-6**	**Ks_z-7**	**Ks_z-8**
Ks_x (cm/s)	0.010	0.010	0.010	0.010
Ks_z (cm/s)	0.00030	0.00050	0.00100	0.010000

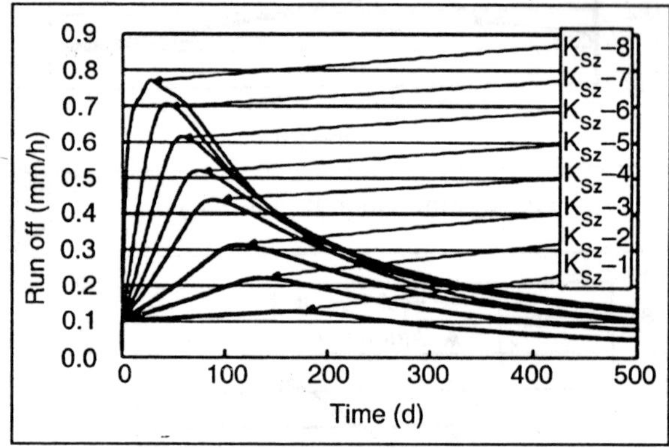

Fig. 4.11 Simulation Result (Ks_z)

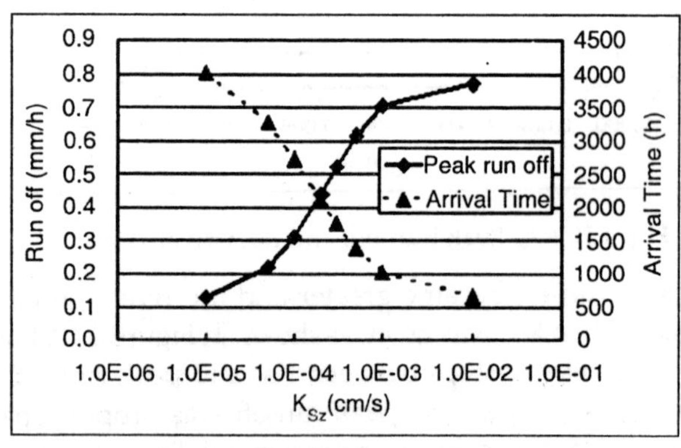

Fig. 4.12 Ks_z, Peak Run-off, and Its Arrival Time

Figure 4.12 shows that peak run-off decreased and arrival times were delayed when Ks_z became smaller. The relationships between Ks_z, peak run-off and the arrival times are shown in Figure 4.11. As Ks_z become greater, peak run-off increased logarithmically and the arrival times decreased exponentially.

Table. Variation of Ks_x and Ks_z

Case No	1	2	3	4	5	6	7	8
Ksx(cm/s)	0.002	0.004	0.006	0.008	0.010	0.010	0.010	0.010
Ksz(cm/s)	0.01000	0.01000	0.01000	0.01000	0.01000	0.00001	0.00010	0.00100
Case No	**9**	**10**	**11**	**12**	**13**	**14**	**15**	**16**
Ksx(cm/s)	0.008	0.006	0.004	0.002	0.008	0.006	0.004	0.002
Case No	**1**	**2**	**3**	**4**	**5**	**6**	**7**	**8**
Ksz(cm/s)	0.00100	0.00100	0.00100	0.00100	0.00010	0.00010	0.00010	0.00010
Case No	**17**	**18**	**19**	**20**	**21**	**22**	**23**	
Ksx(cm/s)	0.008	0.006	0.004	0.012	0.012	0.012	0.012	
Ksz(cm/s)	0.00001	0.00001	0.00001	0.01000	0.00100	0.00010	0.00001	

Twenty-three cases of simulations were performed in order to investigate the relationships between Ks_x, Ks_z, and base-flow. Ks_x and Ks_z were set as shown in Table and the simulation results will be discussed.

CONSIDERATION AND DISCUSSION

On Experiment 1 and Experiment 2

From Figure 4.13 it was obvious that peak run-off quantities of base–flow were related to both Ks_x and Ks_z. However, Ks_x was more effective to the peak run-off quantities for this model, because they varied widely when Ks_x was ranged between 0.002 cm/s to 0.01 cm/s.

On the other hand, it was difficult to choose effective parameter to the arrival time of the peak run-off from Ks_x and Ks_z, because hydrograph became too flat to find out the peak run-off when Ks_x was small though arrival time varied widely with small change of Ks_x comparing to the results of the experiments 2.

The gradients of the recession curves were also changed with Ks_x and Ks_z. It was also obvious that Ks_x would be effective to the gradients of the recession curves.

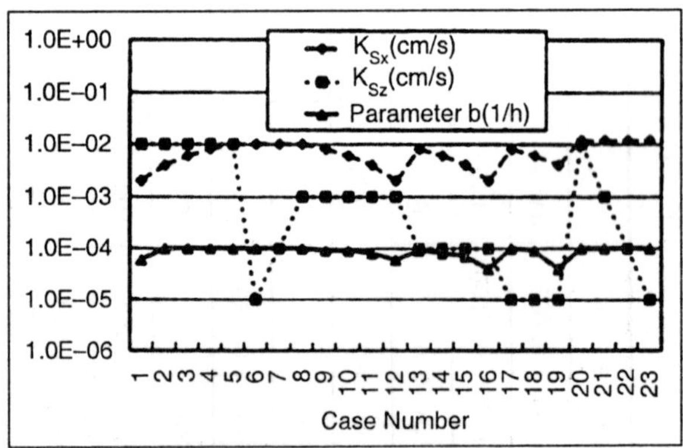

Fig. 4.13 Parameter b and Hydraulic Conductivities (2D)

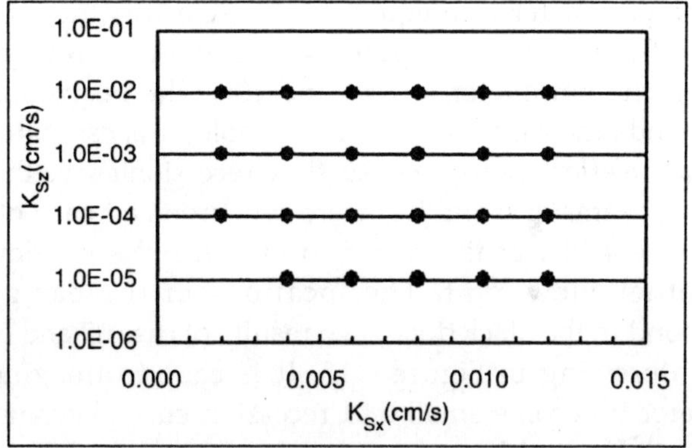

Fig. 4.14 Setting of Ks_x and Ks_z

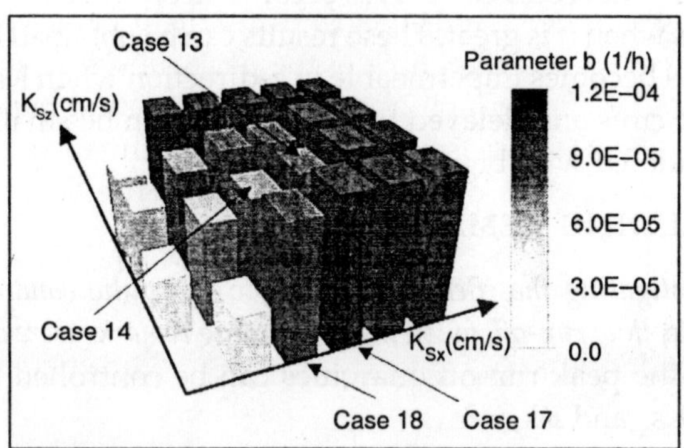

Fig. 4.15 Parameter b and Hydraulic Conductivity (3D)

On Experiments 3

First, each hydrograph was drawn for every case. Then exponential function model, as shown in equation, was applied to the run-off data that were 400 days after the beginning of each simulation and later.

$$q = q_{400} .\exp(-b.t')$$

$$\begin{bmatrix} q:\text{runoff}(mm/h), q_{400} : \text{runoff at 400 days after the begining of the calculation}(mm/h) \\ b:\text{parameter}(1/h), t' : \text{calculation time}(= t-400, \quad t \geq 400)(h) \end{bmatrix}$$

The parameter b in equation characterizes the recession curves of the base–flow. Figure 4.15 shows the relationship between the parameter b, Ks_x, and Ks_z. The horizontal axis corresponds the case number of the Table. It is estimated that Ks_x is more effective to the base–flow recession curve than Ks_z because parameter b and Ks_x were synchronized each other.

Figure 4.15 was illustrated to visualize the relationships from other viewpoint. The locations of the bar graphs correspond to the black dots. Two results of case 17 and case 18 were interesting in Figure 4.15. It is easy to imagine that parameter b become small and recession curve become mild when both Ks_x and Ks_z were small, but parameter b in case 17 and 18 were bigger that those in case 13 and 14, respectively. This means that recession curves become steep when Ks_z is small as well as when it is great. These results can be obtained because the model becomes impermeable in z-direction when Ks_z is less than 10-5 cm/s and delayed time of run-off can be smaller than that calculated with bigger Ks_z.

CONCLUDING REMARKS

Investigating the effect of anisotropic hydraulic conductivities on the base–flow run-off, following concluding remarks were obtained:

- The peak run-off quantities can be controlled by both Ks_x and Ks_z.
- Ks_x is one of the controlling parameters for the characteristics of the long-term base-flow recessions.

All the findings were based on the results of the numerical simulations under restricted conditions, however they can be useful information for the estimations of the "basin-averaged hydraulic conductivity" in the hill-slope direction and its normal directions. The "basin-averaged hydraulic conductivity" can be one of the important input parameters for the constructions of distributed run-off models. The integration of the subsurface run-off model and surface run-off model will allow to evaluate the effects of land-use in future.

5

Economic Development

INTRODUCTION

Beijing is among the biggest cities with serious shortage of water, the amount of water resources per capita is less than 300 m^3, 1/8 as much is per capita in China, and 1/30 in the world. Water resources have become an important factor that conditions Beijing's economic development. It is certain that Beijing will develop sustainablly in the 21st century, but the scale and speed of the development should depend on sustainable utilization of water resources.

CONDITIONED BY WATER RESOURCES

For the past half–century, water conservancy works in Beijing have controlled over 70% of mountainous area, more than 75% of run-off has exploited and utilized and relatively integral system of flood control, water supply, water environment has been established, which strongly supported the social progress and economic development of the capital city. Water crisis, however, occurred many times during the progress, especially, in the five continuous low–flow years in the early of the 1980s, seriously influenced Beijing's economic development.

Rapidly Economic Development with Early Exploitation of Water Sources

Between 1949 and the middle of the 1960s, the exploitation and utilization of water resources in Beijing was in initial stage so that the water resources were relatively abundant and exploited them in time: Guanting and Miyun reservoirs were built successively, using as two big sources of Beijing water supply; Yongding river and Jingmi diversion channels were built as two arteries of Beijing water supply; the 3rd, 4th, 5th, 7th tap water plants using groundwater and the 6th using surface water as sources were built one after another.

Therefore, the development and utilization of water resources accelerated development of society and economy. Industry system of metallurgy, chemistry, electric power, textile and so on were established successively, the gross output value of industry amounted to RMB7.15 billion yuan ($ about 0.86 billion), 40 times as much as the early of the new China foundation, water consumption of industry increased from 30 million m^3 to 640 million m^3 at the average rate of 26% each year.

Moreover, agricultural irrigation area rapidly expanded from 14,000 hm^2 to 246,790 hm^2, and the population of the city was up to 3.75 million. Since the water supply there exceeded the demand for high–speed development, the supply and demand was basically in balance then.

Economic Development Sustained by Excessive Underground- Water Extraction

Heavy industry such as electric power, chemistry, iron and steel industry, of which water consumption is substantial, were built and extended suddenly in Beijing between 1965 and 1980. Gross industrial output value amounted to RMB20.93 billion yuan in 1980, three times as much as in 1965, and meanwhile, industrial water consumption increased to 1.3 billion m^3, twice as much as in 1965.

Moreover, agricultural irrigation area was expanded annually at the same time so that irrigation water reached 2.9

billion m^3, over 60% of total city water consumption. Because of excessive growth of industrial and agricultural production, the contradiction of water supply and demand became increasingly acute and water crisis arose constantly.

In order to obtain top economic effect, Guanting reservoir had to no longer supplied water for agriculture, only for the city and industry.

Because the industrial region in west of Beijing was over concentrated, the water demand could see no way to be met. As a result, people in urban and rural areas had to extracted groundwater competitively, sank 36,134 wells within five years, the amount of annual extraction had reached more than 0.95 billion m^3 that was 50% beyond groundwater recharge in the urban area by the end of the 1970s, and the accumulated deficit amounted to 1.7 billion m^3, which caused groundwater continuously descending by 10 ~20 m, formed a funnel area of 1000 km^2, and led to sources depletion and vegetation regress there. Beijing's development depended on excessive groundwater extraction and the sacrifice of the environment in that stage.

Striving for Development by Water Saving

As Beijing's economy has grown at a high-speed and city scale has extended unceasingly with proceeding of reformation and opening up since 1980s, the industrial and urban domestic water consumption has been rapidly increasing.

However, the exploitable potential is limited, and sources of water supply for city and industry have to turn to original irrigation water. Miyun reservoir, for example, was built mainly for agricultural irrigation, which supplied 0.8~0.9 billion m^3 per year for irrigation before 1980s, but reduced to 0.15 billion m^3 per year in 1994.

Because water shortage became a main factor conditioning Beijing economic growth, the State Council decided that Miyun reservoir would no longer supply water for Tianjing city and Hebei province so as to reduce temporarily water crisis in 1982. During the peak hours of the city water consumption in the middle of the 1980s, however, the amount of water shortage was still up to more than 500,000 m^3 per day, forcing the plants that consume a large quantity of water quitted work, part units of power plant stopped operation, and domestic water of city supplied within a certain time so that the economic growth and social development were influenced seriously by the water shortage.

Because the exploitable sources are very finite, people's attention has to turn to water saving, by which per unit water can create more economic benefits. The rate of recycled industrial water increased from 58% in 1980 to 84.04% in 1992, consequently, the amount of industry water collection that does not include recycled water reduced from1.35 billion m^3 to 0.78 billion m^3, but industrial output value grew 20.93 billion yuan to 59.97 billion yuan, whose average rate of annual growth was 9.2%.

Since irrigation sources are less and less, water-saving-irrigation techniques such as canal lining, low-pressure pipeline, sprinkler and drip irrigation are used, instead of straggling irrigation manner, in order to raise the utilization rate of water. As a result, the grain yield per m^3 of water consumption increased from 0.5 kg to 2.0 kg. The capital development has been sustained by water saving.

PRESENT-SITUATION ANALYSIS OF DEVELOPMENT AND UTILIZATION

Development and Utilization of Water Resources Close to Limit

Development and utilization of water resource has reached high level in Beijing. There 84 reservoirs, 329 sluice gates, 433 small reservoirs and 50 rubber dams, which control over 70% of mountainous area and the rate of water harvesting (p=50%) is up to 75%. For one thing, the exploitation of water resources has approached to limit, Erdaohe reservoir and its division works, planned to build near future, is the only large–scale water conservancy project that will invest RMB1.53 billion yuan, the water supply, however, is only 100 million m3.

For another, excessive extraction of groundwater is very serious. The exploitation in some areas has been beyond the limit of sustainable utilization, causing partially river drying up, groundwater depletion and water-environment aggravation.

Excessive extraction of underground water had accumulated 4.2 billion m^3, and the funnel zone in the urban area had extended to 2100 km^2, 1/8 of total city area. It is impossible to further extend the groundwater extraction, on the contrary, necessary to seek other sources to redress the one so as to rise back in the funnel zone. Consequently, the exploitation potential is very limited.

Inflow in Both Large-Scale Reservoirs Annually Decreasing

The amount of water supply of Miyun and Guanting reservoirs, both large–scale ones in Beijing, was 83% of total surface water supply all over the city in the 1980s. With development and utilization of water resources in the upper outside Beijing, the inflow of the reservoirs sharply reduced from more than 3.0 billion m^3 in the 1950s~1960s to 0.8 billion m^3 in 1995, with a strongly decreasing tendency. Table shows that the annually decreasing tendency of the inflow of Guanting reservoir.

Water-Saving Potential Limited

Since the exploitation of water resources in Beijing has approached to limit, the cost of further exploiting sources is rather high. However, water-saving investment is even lower with better benefits, hence, substantial achievements have been made in Beijing up till now?

The rate of recycled industrial water increased from 48.6% in the 1980s to 86.1% in 1995, only 50% in China in1993 and 75.3% in Japan in 1989, of which the rate of cooling water in Beijing was up to above 90%, therefore, water consumption per ten thousand industrial-output-value reduced from 357m^3 to 65m^3 in1995; irrigation area with water-saving facilities grew from less than 2,600 km^2 in 1980 to 260,000 hm^2 in 1999, 80% of total irrigation area, meantime, the grain yield per m^3 of water consumption increased from 0.5 kg to 2.0 kg.

With increasing development of water-saving engineering, the difficulty and cost are growing; therefore, it is impossible to obtain such benefits like before. Though there is still potential in industrial water saving by means of improving technology and equipments, reformation of technology and equipments simply for water saving is, after all, limited.

It will take long time to renew the technology and equipments to save water. Moreover, irrigation area without water saving in Beijing is only 20% of total irrigation area, it is not, however, ignored that water–saving irrigation certainly reduces redressing groundwater. So water-saving potential is also limited.

Environmental Water Consumption Increasing

Environmental quality is premise of sustainable utilization of water resources and sustainable growth of economy. Since the 1970s, Beijing's water environment has paid a heavy price for its development so that many rivers dry up in most time, even becoming sewage ditch, and play a flood-discharge role only in high-water season in high–flow year, which not only threatens directly people's health, but also contaminates water sources and further aggravates original shortage of water resources. Furthermore, dryly–sandy riverbed becomes sources of sand–dust storm in urban area.

In future planning of the city, the certain amount of water used for green space as well as ecological environment must be considered so as to recover accommodation of rivers, extend urban lake surface, improve environment and climate, prompt environment appreciation, and maintain virtuous cycle of ecological environment. Environmental water consumption, therefore, will certainly increase.

Urban Domestic Water Consumption Rapidly Increasing

Urban domestic consumption that uses for city living is dependent upon two factors, water rating per capita R and population P. The former deals with city character, scale, living level, facilities of water supply and so on.

In a given city, if water rating per capita is R_0 and population is P_0 currently, urban domestic consumption in n years is,

$$W_D(n)=R_0\ (1+a)^n\ P_0\ (1+b)^n$$

Where $W_D(n)$-urban domestic consumption in n years:

α–average annual growth rate of water rating per capita

β–average annual growth rate of population when water rating per capita and population increase annually at the rate of α and β respectively, the urban domestic consumption increases annually at the rate of (α+β+αβ).

Beijing is the capital in China and international metropolis, so the public water consumption of municipal uses and business houses substantially increases with the city–scale expansion; domestic consumption increases with improvement of living level, perfection of water–supply facilities and popularity of domestic water heater; and the population grows also with the city-scale expansion.

As a result, urban domestic water consumption has been increasing without tending to slowdown. which is the most sensitive to water crisis because of requiring concentrative water supply, high rate of guarantee and good water quality. However, domestic water consumption in China, in both urban and rural areas, is only 8% of total water consumption, which does not affect greatly the growth of total water consumption, so it is not taken seriously. In Beijing, however, the proportion of only urban domestic consumption, not including one in rural area, is considerable.

SUSTAINABLE UTILIZATION OF WATER RESOURCES

Gross of water resources W_G is given to a region by nature, basically a constant, while bearing capacity of water resources W_B is ability that meets the demands for man to do social and economic activities under a certain social, economic and technical conditions in a certain stage of social development. Bearing capacity of water resources varies with social and economic development, not only conditioned by gross of water resources, but also affected by man's manner of social and economic activities and water-use efficiency. In given water resources, the higher the development level of society, economy and technology is, the higher the bearing capacity of water resources is.

Generally, bearing capacity of water resources has a marginal value W_{Blim} under a certain economic and technical condition in a given area, which is equivalent to the sum of maximum available water supply W_{Amax} and maximum recycled water W_{Rmax}:

$$W_{Blim} = W_{Amax} + W_{Rmax}$$

The condition of sustainable utilization of water resources is:

$$WA\ ^{3}\ WC \text{ or } WA\ ^{3}\ WU - WR$$

Where: WA- available water supply

- WC–water collection that is consumed directly from surface-water collection and groundwater extraction, not including recycled water
- WU–gross water utilization that is gross used water, including recycled water
- WR–recycled water

Based on the first balance result between water supply and demand in Beijing by 2010, analysed by The Report of Planning Beijing Sustainable Utilization of Water Resources for the Early of the 21st Century, the water shortage will be up to 1.18 (p=50%), 1.65 (p=75%) and 2.00 (p=95%) billion m3respectively, 0.42, 0.66 and 0.93 billion m^3 respectively in the planning urban area. The water shortage is so much that it must restrict the economic development of the capital.

However, it is impossible to take the former measures–simply tapping the potential of water resources–so as to adapt the water supply to the demand of the city economic development.

Before the diversion project from south to north in China (*i.e.* water is diverted across different basins from Yangtze river to Beijing) comes true, Beijing has to continue development and water saving for raising the bearing capacity; and by the local bearing capacity of water, meanwhile, to restrict strategically its scale and speed of the development in the following three respects:

- *Population*: Man is a main body factor of sustainable development, but also brings about negative consequences of resources consumption, especially water resources consumption, and environment destruction at the same time. Population of Beijing, including nonnative and floating population, has amounted to over 15 million currently, increased by about 50% since the middle of the 1980s, which leads to great pressure to the water resources of original scarcity.
- *Scale of the City*: The sudden expansion in city construction that began in the early 1980s with opening up, and still continues nowadays, has enormously increased the need for water. The many new economic development zones have been established in each county and district in Beijing in the past years, of which more than twenty are under construction or have approved. Haidianshanhou in Haidian district and Fengtaihexi in Fengtai district, for example, both development zones near urban area, are under construction. Therefore, the new development zones must take the water resources as the most important constraint condition, developing economy and protecting environment in phase, and doing both water supply and sewage disposal at the same time; otherwise the hidden or delayed water crisis will occur.
- *Speed of Economic Growth*: Speed of economic growth ought to correspond to increase rate of bearing capacity of water resources. Hence, it will certainly

results in new man-made water crisis to overemphasize speed of economic development or to pursue simply economic benefits without consideration of the bearing capacity of water resources. Water shortage in Beijing is a main factor restricting economic growth. The local bearing capacity of water resources must be taken as an important constraint condition planning Beijing future development–act according to actual circumstances, and the scale of the city development and speed of economic growth should be controlled strategically so as to implement sustainable development.

Table. Annually Decreasing Inflow of Guanting Reservoir

Time (Year)	Average Inflow (Billion m³)
1950—1959	1.94
1960—1969	1.30
1970—1979	0.80
1980—1989	0.45
2000 (p=50%)	0.40
2005 (p=50%)	0.30
2010 (p=50%)	0.25

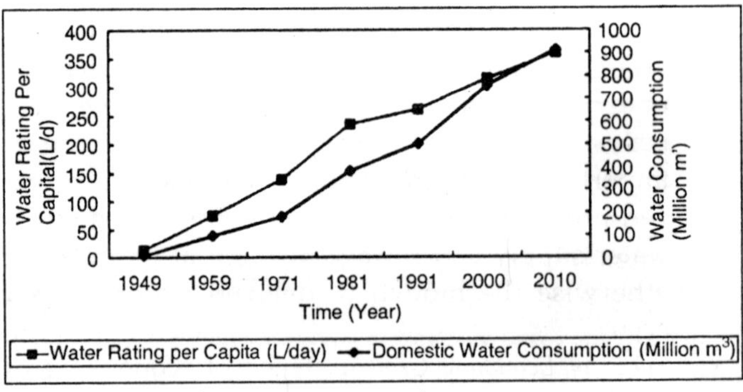

Fig. 5.1 Urban Domestic Water Consumption Increasing Annually

Table. Water Consumption of Domestic, Industry, Agriculture and Environment With Their Percentage of Total Waterconsumption* Billion m³**

Year	Domestic		Industry	
	Water	Percentage	Water	Percentage
1981	0.47	12%	1.10	27%
1995	0.86	22%	1.05	26%
2005	1.17	24%	1.18	24%
2010	1.34	25%	1.35	26%
Year	**Agriculture**		**Environment**	
	Water	**Percentage**	**Water**	**Percentage**
1981	2.47	61%	0.00	0
1995	2.01	51%	0.04	1%
2005	2.08	42%	0.30	6%
2010	2.04	39%	0.35	7%

Table The First Balance Result Between Water Supply and Demand

Area	Item	P=50%	P=75%	P=95%
Whole city	Supply	4.09	3.75	3.40
	Demand	5.27	5.40	5.40
	Balance	-1.18	-1.65	-2.00
Urban area	Supply	1.80	1.57	1.30
	Demand	2.22	2.23	2.23
	Balance	-0.42	-0.66	-0.93

WATER RESOURCE REDUCTION CAUSED BY EXTRAORDINARY

Water management practices including direct measurements such as dams, drainage canals and irrigation system, and indirect measurements such as forest control, affect water quantity and quality in both beneficial and unexpected ways. These beneficial or unexpected effects of management should be evaluated properly. Reduction of water resource in a watershed caused by forest growth and canopy amassment has not been well noticed, although beneficial functions of flood reduction and soil conservation are widely known.

Forest is vitally necessary to maintain the ecosystem in a watershed, to adjust stormwater before forming dangerous flood peaks, to prevent soil erosion on slopes, and to give productive timbers. A watershed without forest is vulnerable to storm rains and debris flow. However, too much forests or extraordinarily planted trees are not the ideal circumstance, because in that case the evaporative water loss from foliage and canopy might be very large, and the run-off in river might get decreased.

And any kind of forest can not substitute the functions of dam for cutting flood peak and supplying water in drought seasons. Therefore there should be an optimistic range for forest foliage (or biomass): below this range the watershed is dangerous with floods and debris, above this range the watershed is possibly in shortage of water resource. This paper does not identify the optimistic range of forest, but aims to demonstrate the minus or negative effects (not yet well noticed) caused by forest increase and growth. Past data of water balance in two watersheds in Japan are analysed and used to demonstrate the unexpected reduction of water yield, as Japan is typical of rapid plantation. Yearly changes of annual water budgets are found from field data, and the reasons for water reduction are explained.

RESULTS AT NAGAYASUGUCHI BASIN

The basin of Nagayasuguchi Dam is the upper stream catchment of the Naka River in Tokushima Prefecture, Shikoku Island of Japan, with an area of 494 km^2, with rich precipitation, and being covered by conifer and broadleaf trees. Based on data of 37 years of annual rainfall and run-off, annual evapotranspiration amount can be simply estimated by water balance principle (annual evaporation is the residual of rainfall minus run-off). Water budgets of each year and their yearly changing trends. The 10–year moving mean values are calculated, and are used to analyse the changing trends.

Two periods of ten years, 1957-1966 and 1984-1993, will be considered for comparisons as shown in Table. Annual rainfall had a small increase trend within the 37 years, with the 10-year mean value increased by 165 mm (or 5.1%) from 3,204 mm for 1957-1966 to 3,369 mm for 1984 –1993. On contrast, annual water yield (run-off) had a tiny decrease trend, the 10-year mean of run-off decreased by 68 mm (2.5%) from 2,755 mm in 1957-1966 to 2,687 mm in 1984 overemphasize –1993. Between the same periods, evapotranspiration showed a dramatically evident increase trend, increased by 308 mm (60.8%) from 506 mm to 814 mm. As a result, annual run-off got decreased although rainfall was increasing, because the evaporative loss increased extraordinarily.

Table. Comparison of Water Budgets of Two Periods for the Nagayasuguchi Dam Basin

	1984-1993	1957-1966	Change (mm)	Percentage (%)
Rainfall (mm)	3369	3204	+165	+5.1
Runoff (mm)	2687	2755	-68	-2.5
Evapotranspiration (mm)	814	506	+308	+60.8

Then why did the evapotranspiration increase so much during the 37 years? Possible reasons should include changes and effects of rainfall, temperature, forest, agricultural activity and domestic water usage. Effects of rainfall on evapotranspiration are complex, depending on intensities of rain events and rainy days in a year. It is roughly supposed that the contributing fraction to evapotranspiration of the increased rainfall (165 mm) be same as the evaporation-rainfall ratio: 814/3369 or 0.242. That means the rainfall increase contributed about 40 mm to evapotranspiration increase, giving only 13% contribution to the total 308 mm of evapotranspiration increase.

Alteration of temperature causes the potential evaporation to change, and then affects actual evapotranspiration. Mean temperature decreased from 14.88? in 1957-1966 to 13.19? in 1984-1993, and potential evaporation decreased by 9.22% if the potential is estimated with the Hamon formula. Its effect to actual evapotranspiration is roughly supposed to be a decrease in this percentage, *e.g.* a minus contribution of 47 mm (506 9.22%). That means the real or net increase of evapotranspiration is 355 mm (308+47).

Agricultural activities and domestic water usage did not change evidently, because the basin is a natural mountain area with little resident people. However, there was a rapid growth in area of forests and canopy's foliage mass during 1960 through 1990, caused by the policy of extending plantation. The tree density has arisen to three times of that in 1950s, canopy foliage has become doubled. This increase of trees and foliage mass produces more canopy interception of rainfall, takes more soil water for transpiration. Therefore among the net evaporation increase of 355 mm, besides the 40 mm due to rainfall increase, the residual 315 mm is most possibly caused by the forest plantation and growth. Forest change is the main reason for evaporation increase. Some theoretical and experimental researches on the function of forest foliage in evapotranspiration processes were made by the authors and other researchers.

Roughly the interception evaporation and transpiration are positively proportional to the foliage area and volume. When canopy biomass gets larger, more rainfall water can be intercepted and evaporated, more soil water may be transpirated. These phenomena can not be explained by only energy balance theory, but should be explained by complex three-dimensional thermal-dynamic and aerodynamic theories which are not the main topics of this paper.

RESULTS AT HIJI BASIN

As for the basin of Hiji River in Ehime Prefecture of Shikoku, with an area of 1,009 km^2 and averaged rainfall of 1,800mm, a similar analysis of annual water balance is made for 36 years of 1960–1995. A 85% of this basin is covered by forest and has experienced an increase in trees' foliage. A typical forest stand is shown in Fig. 5.1 and the high density can be seen. Basin averaged rainfall is obtained by a spline interpolation algorithm. Annual evapotranspiration is the subtraction of run-off from rainfall. Annual budgets and 10-year moving means are shown in Fig.5.1 Comparing the 10–year means of water budget items between the period 1960–1969 and 1983–1992, as listed in Table, it is found that annual rainfall increased by 34 mm (2.0%) from 1,721 to 1,755 mm, but run-off decreased by 140 mm (11.1%) from 1,258 to 1,118 mm, while evapotranspiration increased by 173 mm (37.3%) from 464 to 637 mm, showing a similar change pattern as in the Nagayasuguchi Dam basin.

Table. Comparison of Water Budgets of Two Periods for the Hiji River Basin

	1983-1992	1960-1969	Change (mm)	Percentage (%)
Rainfall (mm)	1755	1721	+34	+2.0
Runoff (mm)	1118	1258	-140	-11.1
Evapotranspiratio n (mm)	637	464	+173	+37.3

Supposedly, increased rainfall (34 mm) contributed 12 mm to evapotranspiration according to the evaporation-rainfall ratio of 0.363, possessing only 6.9% in the total 173 mm increase of evapotranspiration. The change of rainfall can not interpret the change of evapotranspiration. On the other hand, mean temperature had a tiny decrease: from 13.2°C in 1960–1969 to 12.8°C in 1983–1992, causing the potential evapotranspiration to decrease by 2.26%, or causing actual evapotranspiration to decrease by 11 mm. As a result, rainfall–caused increase amount of 12 mm is almost deleted by the potential evapotranspiration-caused decrease amount of 11 mm.

Domestic water usage and its alteration are negligible for this basin, because farmland possesses a small part of the basin. Therefore, forest foliage increase is identified to be the main reason for the 173 mm increase of evapotranspiration. In fact, canopy foliage mass has arisen by about 30% according to forest investigation statistics. The changes of tree age can be used to interpret the increase in tree's biomass. Averaged tree age and its changes during 30 years are listed in Table. For both the conifers and broadleaf trees, their age was increasing yearly. That means the amount of tree growth and plantation is larger than the amount of tree death and felling. If the older trees had been cut down or felled while the young trees were growing, the averaged tree age would not have increased as the table shows.

In fact, plantation was largely implemented during 1960s and 1970s based on the national plantation policy, but matured stands were not properly felled for timber usage because of rapid extension of import of foreign timber and because of decrease in forestry labors. As a result, the forest in the basin was not carefully managed, and tree density and canopy biomass became very large. Then evaporative loss of water has also substantially increased.

Table. Changes of Tree Age in the Hiji River Basin

Years group	Conifer trees		Broadleaf trees	
	Area (ha)	Age (year)	Area (ha)	Age (year)
1962-63	46,284	18.5	23,281	15.0
1967-68	49,920	19.6	19,024	17.2
1972-73	53,802	21.0	15,740	18.3
1977-78	54,219	23.7	14,958	20.4
1982-83	50,996	25.2	18,354	22.2
1987-88	51,604	28.0	17,411	24.2
1992-93	51,619	31.6	17,386	27.8

Annual water budgets and their yearly changes were obtained from field observatory data in two basins of Shikoku, Japan. They showed a similar changing pattern: a tiny increase in rainfall, a contrary decrease in run-off, and a dramatic increase in evapotranspiration loss. This phenomenon is explained to be caused by extraordinary growth and increase of forest foliage, as increased foliage resulted in larger interception and transpiration. Therefore, apart from many expected benefits such as floods reduction and timber productivity, extra forest growth or plantation may cause the unexpected effect of reducing water yield. This phenomenon should be notified especially for those regions with water scarcity.

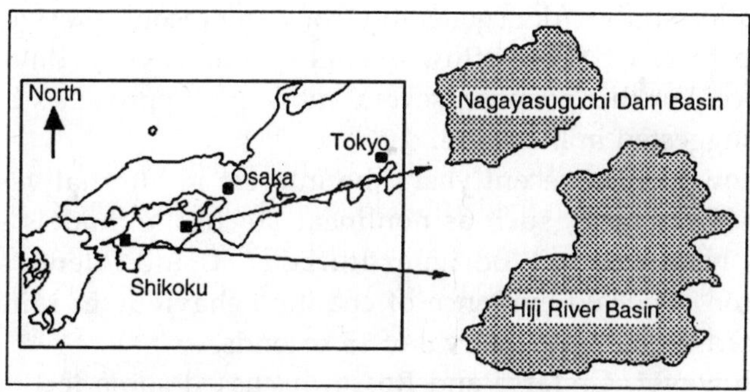

Fig. 5.2 Locations of the Nagayasuguchi Dam Basin and Hiji River Basin

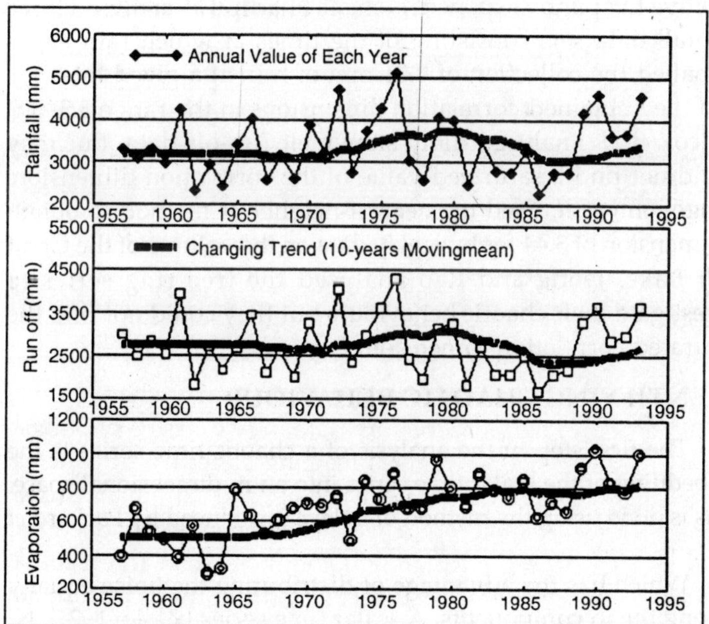

Fig. 5.3 Water Budgets and Changing Trends at the Nagayasuguchi Basin

CHAOS AND SAMPLED DAILY STREAMFLOWS

The modeling of daily streamflow time series has attracted the attention of hydrologists for a long time. The main reason for such interest has been the need to conduct synthetic simulation studies (data generation) of water resources systems, and to forecast future flow events one or several days in advance. For this purpose, several modeling approaches have been suggested in literature.

However, in recent years the interest in alternatives to stochastic models, such as nonlinear stochastic models and chaos, has increased. odriguez-Iturbe *et al.* and Puente and Obregon reported evidence of chaotic behaviour in rainfall recorded with a time interval of 15 seconds.

However, Ghilardi and Rosso discussed some technical issues involving data size, and they pointed out that it is difficult to discriminate between chaos and noise when the largest positive Lyapunov exponent is small. Sharifi *et al.* analysed three rainfall data sets consisting of the times at which rain gages signalled the collection of 0.01 mm of rain at a given location, and they obtained correlation dimensions in the range 3.3–3.8. Wilcox *et al.* analysed daily snowmelt run-off data, but they could not find a saturated value of the correlation dimension. Sangoyomi *et al.* used the nearest-neighbour method to obtain a dimension of 3.44 for biweekly data on the volume of the Great Salt Lake. Jeong and Rao analysed the tree ring series to investigate their chaotic behaviours but they could not find the saturated correlation dimension.

IDENTIFYING CHAOTIC BEHAVIOUR

The first step in the analysis of a chaotic time series is the embedding of the scalar time series into an m-dimensional space. This is done using the method of delays introduced by Packard *et al.*

Which has the advantage of distributing the noise equally among the m components. A scalar time series $\{x_i\}$, i = 1, 2..., N, is embedded into m-dimensional space by constructing the vectors.

Correlation Integral and Correlation Dimension

After the attractor has been reconstructed using Eq. quantitative properties of the chaotic system can be determined. The cor relation dimension introduced by Grassberger and Procaccia is widely used in many fields for the quantitative characterization of strange attractors. The correlation integral for the embedded time series is the following function:

$$C(m,N,r,t)=\frac{2}{M(M-1)}\sum_{1\leq i<j\leq M}\Theta\left(r-\left\|\vec{X}_i-\vec{X}_j\right\|\right),\quad \gamma>0$$

where $\Theta(\alpha) = 0$, if $\alpha < 0$, $\Theta(\alpha) = 1$, if $\alpha \geq 0$,

N is the size of the data set, $M = N - (m - 1)t$ is the number of embedded points in m-dimensional space, and $\|....\|$ denotes the sup-norm. C(m,N,r,t) measures the fraction of the pairs of points $\vec{X}_i$, $i = 1,2..., M$, whose sup-norm separation is no greater than r. If the limit of C(m,N,r,t) as $N\to\infty$ exists for each r, we write the fraction of all state vector points that are within r of each other as $C(m,r,t) = \lim_{N\to\infty} C(m,N,r,t)$, and the correlation dimension is defined as,

$$D_2(m,r,t) = \lim_{r\to 0}\left[\log C(m,,r,t)/\log r\right].$$

In practice, N remains finite, and, thus, r cannot go to zero; instead, we look for a linear region of slope $D_2(m,t)$ in the plot of log C(m,N,r,t) vs. log r.

The Parameters M and T

The components of the reconstructed state variables $\vec{X}_i$ need to be independent, so the quality of the reconstructed attractor depends on the choice of the index lag t.

If the delay time τ_d is too small, the reconstructed attractor is compressed along the identity line, and this is called redundance. If τ_d is too large, the attractor dynamics may become casually disconnected, which is called irrelevance, and which may cause the attractor to appear much more complex than it really.

Choosing the Delay Time and Delay Time Window

Many researchers choose to use a fixed delay time τ_d as the embedding dimension m is increased. Some have suggested obtaining τ_d from the autocorrelation function (ACF), which is practically convenient, since it contains information about both periodic trends and information dissipation. This ACF method has the advantage of computational efficiency, but it has been found that the value obtained for t may be incorrect.

Since the relationship between the spatial distribution of a reconstructed attractor and the temporal autocorrelation of a single-variable time series is not well-defined, there are inconsistencies inherent in this approach instead suggested choosing the index lag t as the first local minimum of the mutual information (MI).

It is known that this is the most comprehensive method, but it has the drawbacks that it requires a large amount of data, and it is cumbersome computationally.

We mentioned the alternative of fixing the delay time window $\tau_{w,}$ rather than the delay time $\tau_{d,}$ but the estimation of τ_w is less well developed. Martinerie *et al.* examined the delay time window and compared it with the delay times estimated using the ACF and the MI. They concluded that τ_w could not be estimated using either of these two methods. Basically, τ_w is the optimal time for independence of the data, but these methods estimate the first locally optimal time, which is $\tau_{d.}$ From this distinction between τ_d and $\tau_{w,}$ we developed a technique, called the C-C method, that can estimate both τ_d and $\tau_{w.}$

This method is discussed in the following subsection. We also showed that, for small data sets, as the embedding dimension m is increased, the correlation dimension D_2 converges more rapidly if τ_w is held fixed than if τ_d is held fixed.

The C-C Method

Brock *et al.* studied the BDS statistic, which is based on the correlation integral, to test the null hypothesis that the data are independently and identically distributed (iid). This test has been particularly useful for chaotic systems and nonlinear stochastic systems. Under the iid hypothesis, the BDS statistic form > 1 is defined as,

$$BDS(m,M,r) = \frac{\sqrt{M}}{\sigma}\left[C(m,M,r) - C^m(I,M,r)\right],$$

and this converges to a standard normal distribution as M→ ∞ Note that the asymptotic variance σ^2 (m, M, r) can be estimated as,

$$\sigma^2(m,M,r) = 4\Big\{m(m-1)C^{2(m-1)}\left(K - C^2\right) + K^m - C^{2m}$$
$$+2\sum_{i-1}^{m-1}\left[C^{2i}\left(K^{m-i} - C^{2(m-i)}\right) - mC^{2(m-i)}\left(K - C^2\right)\right]\Big\}$$

$$C(m,M,r) = \frac{2}{M(M-1)}\sum_{1 \le i < j \le M}\Theta\left(r - \left\|\bar{X}_i - \bar{X}_j\right\|\right),$$

$$K(m,M,r) = \frac{6}{M(M-1)(M-2)}\sum_{1 \le i < j < K \le M}\Theta\left(r - \left\|\bar{X}_i - \bar{X}_j\right\|\right)\Theta\left(r - \left\|\bar{X}_j - \bar{X}_k\right\|\right).$$

The present study is concerned with the properties of the quantity S(m,n,r,t) = C(m,N,r,t)–C^m(I,N,r,t). We refer to a comment by Brock *et al.* " If the stochastic process { X_i } is iid, it will be shown that C(m,r)–C^m(I,r) for all m and r. That is to say, the correlation integral behaves much like the characteristic function of a serial string in that the correlation integral of a serial string of independent random variables is the product of the correlation integrals of component substrings." This led us to interpret the statistic S(m,N,r,t) as the serial correlation of a nonlinear time series. Therefore, it can be regarded as a dimensionless measure of nonlinear dependence. For fixed m, N, and r, the plot of S(m,N,r,t) vs. t is a nonlinear analog of the plot of the autocorrelation function vs. t. In order to study the nonlinear dependence Kim et al derived the following equations:

$$S(m,r,t) = \frac{1}{t}\sum_{5-1}^{t}\left[C_5(m,r,t) - C_5^m(1,r,t)\right], \; m = 2, 3...$$

$$\Delta S(m,t) = \max\{S(m,r_j,t)\} - \min\{S(m,r_j,t)\}.$$

Brock et al suggested that m should be between 2 and 5, and r should be between $\sigma/2$ and 2σ. In addition, the asymptotic distributions were well approximated by finite time series when $N \geq 500$.

Thus, we select four values of r in the range $\sigma/2 \leq r \leq 2\sigma$, $r_1=(0.5)\,\sigma, r_2=(1.0)\,\sigma$, $r_3=(1.5)\,\sigma$, and $r_4=(2.0)\,\sigma$, as representative values. We then define the following averages of the quantities given by Eqs.

$$\bar{S}(t)=\frac{1}{16}\sum_{m-2}^{5}\sum_{j-1}^{4}S(m,r_j,t),$$

$$\Delta\bar{S}(t)=\frac{1}{4}\sum_{m-2}^{5}\Delta S(m,t),$$

and we look for the first zero crossing of $\bar{S}(t)$ or the first local minimum of $\Delta\bar{S}(t)$ to find the first locally optimal time for independence of the data, which gives the delay time $\tau_d = t\tau_5$. The optimal time is the index lag t for which $\bar{S}(t)$ and $\Delta\bar{S}(t)$ are both closest to zero. If we assign equal importance to these two quantities, then we may simply look for the minimum of the quantity,

$$S_{cor}(t)=\Delta\bar{S}(t)+\left|\bar{S}(t)\right|,$$

and this optimal time gives the delay time window $\tau_w = t\tau_5$.

APPLICATIONS TO DAILY STREAMFLOW DATA

In this work, we search for evidence of chaotic behaviour in the daily streamflows of two different stations: St. Marys river near Macclenny and Ocklawaha river near Conner, Florida, USA.

The two streamflow records consist of 67 years with 24,472 daily data for St. Marys river near Macclenny and 11 years with 4,018 daily data for Ocklawaha river near Conner. The time series plots for these streamflows are shown in Fig 5.3.

Estimations of τ_d and τ_w

For the streamflows of St. Marys and Ocklawaha rivers, τ_d corresponds to the first local minimum points of $\Delta\bar{S}(t)$, as indicated by the arrow in Fig. for the case of St. Marys river and it can be determined similarly for Ocklawaha river. τ_d on both streamflows are estimated as 39 τ_5 =39 days and 33 τ_5 =33 days.

Also, as shown in Fig. 5.2, we found that the minimum of S_{cor} (t) occurs for t = 152, which gives τ_w = 152 days for St. Marys river and τ_w = 187 for Ocklawaha river.

Correlation Dimensions For Daily Streamflows

Fig 5.4 shows the plots of log[C(r)] versus log(r) for the reconstructed attractors for St. Marys river for embedding dimensions m = 2, 4..., 20 using τ_d = 39 days.

The correlation dimension slowly increases up to m = 20 and there is no evidence for chaotic behaviour. In Fig. 5.4, we repeat this analysis using the delay time window τ_w = 152 days. Since the index lag is then determined by (m-1)t=152, it is necessary to round off t to the nearest integer. In this case, also, we can not find the correlation dimension which describes the chaotic behaviour.

A similar analysis for Ocklawaha river is performed but, in both cases of τ_d and $\tau_{w,}$ the correlation fails to saturate as shown in Fig 5.3, so there is also no evidence of nonlinear determinism.

SAMPLING OF CHAOTIC TIME SERIES

Sampling is an important consideration in the analysis of hydrologic time series. Hydrological processes such as precipitation and streamflow are generally continuous in time. While continuous time series are measured at some gaging stations, most recorded time series are discrete, providing instantaneous sampling at either regular or irregular time intervals (such as instantaneous daily observations of water levels in streams).

As an example, we examine the effect of sampling on a time series of the variable x from the Lorenz equations:

$$\dot{X} = -\alpha(x - y)$$

$$\dot{Y} = -xz + cx - y$$

$$\dot{Z} = xy - bz$$

which is generated using the parameter values a = 16.0, b = 4.0, and c = 45.92 and a time step of $\tau_5 = 0.01$. The sampled time series will consist of 15,000 values. We generate sampled time series by keeping only the values at every nth time step, with n=2, 10, 50, and 100. For each of the four sampled time series, we use the C-C method to compute the delay time $\tau_{d,}$ and the results are given in Table. We then construct embeddings of the attractors using these delay times.

The attractors degrade quickly as n increases, and, by n = 10, they are completely unrecognizable. We also compute the correlation dimensions, and these results are shown. In each case, we draw a horizontal line at the dimension $D_2 = 2.05$ obtained from the original time series. Note that, for the larger values of n, it is not always possible to find a linear region in the plot of log [C(r)] versus log (r) for large values of the embedding dimension m. The convergence of the correlation dimension degrades slowly as n increases, and, by n = 100, this convergence is lost. Thus, we see that sampling can eliminate the evidence of nonlinear determinism from a chaotic time series.

Table. The Delay Times and the Delay Time Windows for Sampled Lorenz Series.

Time Interval	τ_d	τ_w	Time Interval	τ_d	τ_w
2	5	134	50	2	26
10	2	91	100	2	24

Since, as the time interval n is increased, the time series is more and more randomized by the sampling process, to find the best optimal point of S_{cor} (t) for τ_w is difficult. Thus, we have not used the delay time window for the estimation of the correlation dimension in sampling process.

DISCUSSIONS AND CONCLUSIONS

The Lorenz equations were solved numerically using a time step of $\tau_5 = 0.01$, and the resulting values of the variable x were used as a scalar time series. Applying the C-C method to this time series yields the delay time $\tau_d = 10\ \tau_5 = 0.1$ and the delay time window $\tau_w = 100\ \tau_5 = 1.0$. We then investigated the effect of sampling on this time series by keeping only the values at every nth time step. This is equivalent to making sampled measurement with a regular time interval $\tau_m = n\ \tau_5$. As this time interval τ_m increases, the successive measurements will eventually become irrelevant, and any nonlinear determinism will be lost. When this happens, the data will appear to be stochastic, rather than chaotic.

Since the delay time τ_d is a basic correlation time, then irrelevance should begin to become apparent when $\tau_m \sim \tau_d$. However, τ_w is the maximum time for correlations, so complete stochasticity should not occur until $\tau_n \sim \tau_w$. For the Lorenz system, the conditions $\tau_m \sim \tau_d$ and $\tau_m \sim \tau_d$ become n~10 and n~100, respectively. Stochastic models are often used for the modeling of daily streamflows. However, if the streamflow data shows evidence of nonlinear determinism, then a chaotic model should be more appropriate. Some previous studies have found evidence of nonlinear determinism, but some studies have not found such evidence, may be, due to sampling process. Therefore, one should not expect to find evidence of nonlinear determinism in this case, and, stochastic models should work well.

CHARACTERIZATION OF GROUNDWATER

The emerging problem of groundwater contamination due to agricultural practices and increased use of fertilizers and pesticides has created a need for information on groundwater quality. Corresponding awareness of the importance of groundwater quality monitoring has also become a focus of attention. In order to extract the information, which regulatory agencies and other organizations need for effective management of the groundwater resources, statistical analysis of monitoring data is imperative.

The design of effective monitoring (statistical sampling) programmes and the selection of appropriate statistical methods for analyzing groundwater quality variables requires an understanding of the behaviour of the random variables of concern. Statistical analysis of data is performed in order to detect the changes in groundwater quality over time or over space.

The choice of statistical methods is dictated both by the information expectations of water quality regulators and the statistical characteristics of the water quality variables. From a statistical viewpoint, all water quality constituents are considered random variables.

The effective design of monitoring programmes and subsequent utilization of data obtained depend upon an understanding of the general statistical characteristics of groundwater quality variables. Moreover, if suitable probability distribution function can be fitted to observed data then the values needed by the hydrologists and engineers may be extrapolated beyond the range of the observational period.

To select appropriate statistical tests for change, one must know whether the water quality variables of concern:

- Are normally distributed,
- Exhibit seasonal variations and
- Are correlated in time or serially dependent.

In monitoring and evaluation of the effects of the contamination source of groundwater, information on the spatial variation in the background groundwater quality is also needed.

Characterization of variables in these terms is a necessary first step in regulatory data analysis and hence it was performed in this study. This step of analyses ensures the validity of assumptions on which the selected trend testing methods are based.

STUDY AREA AND DATA COLLECTION

Northwestern zone of Bangladesh was selected as the study area that includes the greater administrative districts of Dinajpur, Rangpur, Rajshahi, Bogra and Pabna. The area is situated between longitudes 88°10'and 89°50' E and between latitudes 23°30'and 26°38'N. The region is adjacent to the Indian states of Bihar and West Bengal and forms a hydrological distinct area bounded on the south by the Ganges-Padma river and on the east by the Brahmaputra-Jamuna river.

The major land types in the region include recent alluvial flood plains and Pleistocene terraces. Abundant groundwater supplies occur throughout the region in a shallow unconfined aquifer. The northwestern zone was selected for this study because the number of tubewells, the level of groundwater use and the cropping intensity were found to be significant in this area. Groundwater quality data were collected from Bangladesh Water Development Board (BWDB)-a regulatory agency responsible for nationwide groundwater quality monitoring (BWDB, 1976-84). BWDB collects water samples from 117 tubewells and piezometric wells throughout the year in whole Bangladesh. Of these, approximately 40 designated observation wells are located in the northwestern zone. Nineteen different parameters for groundwater quality are usually determined by BWDB. Among them, concentration data on nine parameters such as, chloride (Cl), nitrate (NO_3), total dissolved solid (TDS), pH, sodium adsorption ratio (SAR), sodium (Na), iron (Fe), calcium (Ca) and magnesium (Mg) were selected for analyses. These parameters were considered because most of them are likely to show the changes in quality due to variations in infiltrations over time. Data series of 1973-1983 for these parameters were selected for analysis because of its complete database.

DATA ANALYSIS AND DISCUSSION

Normality

Of all the commonly used probability distributions, normal distribution is most widely used because of its performance as a base distribution for comparison and error analysis. Large departures from normality, particularly in the form of skewness, or lack of symmetry, can invalidate results. Transformations (for example, log transformation) of non-normal data are often used to remove skewness and produce normally distributed values. In this study, data were examined for normality by several alternative methods, such as fitting empirical distribution, skewness test and chi–square goodness–of–fit test.

Empirical Distribution

Empirical distribution of observations is usually represented by frequency histograms, which provides a visual indication of the symmetry of probability distributions.

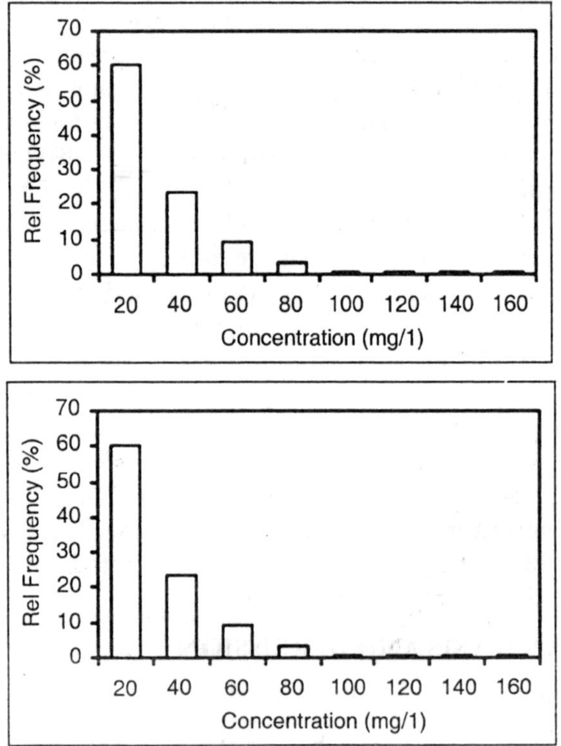

Fig. 5.4 Frequency Histogram for Chloride (Cl) Data Series

Histograms were constructed for original and logtransformed data sets for all the parameters. Histogram for chloride is presented in Fig. 5.4 in order to perform a visual inspection for normality.

Relative frequency provides an estimate of the probability of parameter concentration falling in the indicated range or class interval. Frequency distribution of all the parameters appeared to be skewed to the right.

Qualitatively it can be stated that the degree of skewness varied considerably for the parameters, which may contain only a few values that are significantly larger than the average values. These values may arise from measurement errors or from groundwater contamination, in which case the high values may belong to a "population" different from that of the remaining sample values. However, the groundwater quality variables were found lognormally distributed, which is similar to Chow *et al.*.

Skewness Test

Skewness is a measure of symmetry of the distributions and can be a conclusive indicator of non-normality. The skewness co-efficients (Cs) for original and logtransformed data were calculated following the procedure given.

Results revealed that skewness of all the parameters have positive value except pH. Negative skewness is not as common as positive skewness in groundwater quality, but the example is affected by a very few sample points which are very much lower than the rest.

Theoretical values of skewness for normal and lognormal distributions are reported by Law and Kelton as 0.00 and 6.18 respectively. Comparing the computed values with these values for normal and lognormal distribution, all water quality parameters were appeared to be lognormally distributed.

However, most of the skewness for logtransformed data shows negative value, which is usually expected for nonnormal data set.

Chi-Square

The Chi-square goodness of fit test was used to test for a significant difference between the distribution suggested by a data sample and a selected probability distribution. Here the test assumed the data drawn from a normal population; chi-square test checked the validity of this assumption.

The hypothesis of χ^2–test was performed at 5%significance level and the computed χ^2 was compared with critical χ^2 value. The results revealed that the computed χ^2 values for the groundwater quality parameters are larger than the critical χ^2 values, which indicates that the parameters were not normally distributed.

Seasonality

Hydrogeologic conditions of the site may suggest specific forms of seasonality, which can be tested. For example, with quarterly data, a reasonable form of annual cycle of one quarter may be different from the other three. Again high recharge occurs in the monsoon season.

So form of annual cycle in the monsoon season (high recharge period) may be different from other seasons. If no prior determination of seasonality can be made, each of the seasons must be tested to determine whether its mean tends to be different from that of the other seasons. Prior to seasonality test, it is necessary to adjust data series to make it stationary, which can be obtained by transforming data into zero-mean or trend-free series. For this, the data was organized into quarterly values; as such, the first quarter of a year was defined as January, February and March.

Zero-mean or trend–free series was obtained in the following manner: a linear regression of concentration versus time was plotted and the slope of the regression was tested for significance (at 5%).

Those parameter values showed significant linearity, the trend (a+bt) was removed from the sample values, otherwise, the mean concentration was removed from the parameter observation and thus, a series with trend free or zero-mean was obtained. Following this procedure, mean was removed from Cl, NO_3, TDS, SAR, Fe, and trend was removed from pH, Na, Ca, Mg data and thus the series was made stationary. The stationary data were then checked for seasonality by different statistical tests such as, 2–sample test

In the analysis quarter 1, and 2 were taken as group 1 and quarter 3 and 4 were taken as group 2 in 2– sample tests. Quarter 1, 2, 3 and 4 were taken as group 1, 2, 3 and 4, respectively for 4–sample test.

Summary of all these test results are given in Table. In 2-sample test, NO_3, TDS, SAR and Mg show seasonality in t-test and Mann-Whitney U Test respectively. The proportion of NO_3 ions originating from fertilizers and other agricultural practices varies from season to season.

SAR and Mg of group 1 (dry and premonsoon season) should be higher than the group 2 (monsoon and post monsoon season) due to different in precipitation. Again TDS (ppm) is a measure of salinity that can be expected to increase through the combined effects of groundwater recycling, high evapotranspiration and low rainfall.

During dry period agriculture depends more on groundwater pumping conceivably, the water recharged in this period is recycled and concentrated due to evapotranspiration, resulting in an increase of groundwater salinity. Students t-test shows seasonality at higher level of significance (20%) because that the test assumes the data are independent and come from normally distributed populations with equal variances, which was not true for the present case.

It was seen in the 4–sample tests that Cl concentration shows seasonality. That is, mean Cl concentration differs from season to season. From ANOVA, it was further seen that the mean of group 3 differs from the means of group 2 and 4 at 5% significance level. Group 3 indicates the monsoon season of July, August and September with high precipitation and recharge events and thus, Cl concentration may be assumed to be lower in this season.

Table. Summary of Statistical Tests to Detect Seasonality at Different Level of Significance (%)

Variables	2-sample tests		4-sample tests	
	t-test (20%)	Mann-Whitney (5%)	ANOVA (5%)	Kruskal-Wallis (5%)
Cl	No	No	Yes	Yes
NO_3	Yes	Yes	No	No
TDS	Yes	Yes	No	No
pH	No	No	No	No
SAR	Yes	No	No	No
Na	No	No	No	No
Fe	No	No	No	No
Ca	No	No	No	No
Mg	No	Yes	No	No

Serial Dependence

Serial correlation can exist between an observation at one time period and an observation k time periods earlier for k = 1, 2.... In this discussion of serial correlation, it is assumed that observations are equally spaced in time (quarterly sampling frequency). The population serial correlation coefficient is denoted by p(k) (frequently called the autocorrelation coefficient) where k is the lag or number of time intervals between the observations being considered. For water quality processes, the function p(k) decays with increasing k.

The sample lag–k autocorrelation coefficient for a sample of size n was calculated using the process given by Haan. The sample serial correlation coefficient is denoted by r(k) which was tested for significance using confidence interval approach. As autocorrelation coefficients range from -1.0 to +1.0, where positive values indicate a direct relationship.

For example, a positive lag one autocorrelation coefficient r(1) would mean that high values in one time period tend to be followed by high values in the next time period. Negative autocorrelation implies an inverse relationship. Autocorrelation coefficients (for k = 1, 2... 8) for all the parameters were calculated and tested for significance at 95% confidence intervals. Among them, autocorrelation function of Cl and Na is shown in Fig 5.5.

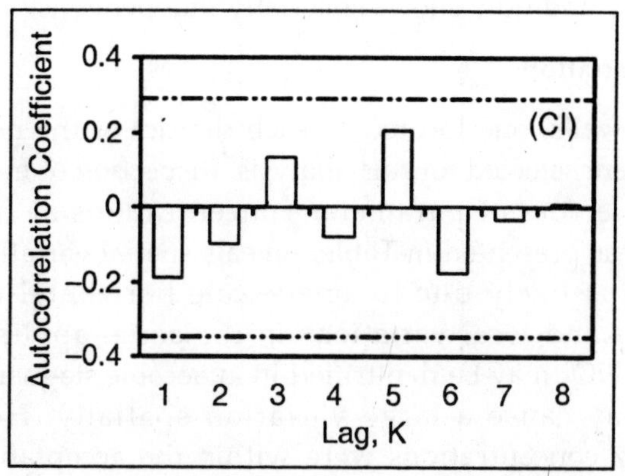

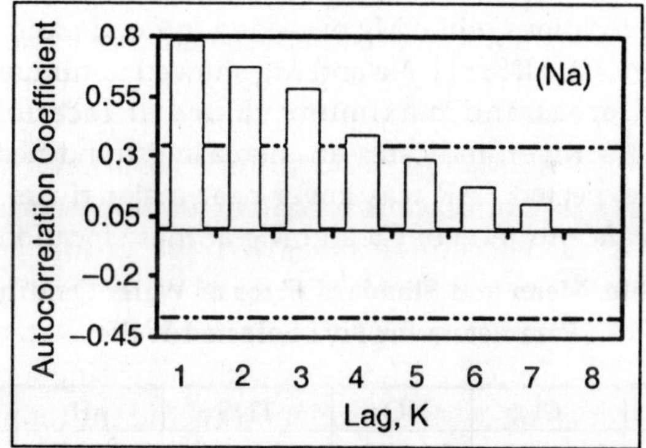

Fig. 5.5 Auto Correlation Function Exhibited by Cl and Na (Sample Frequency of Data). Dashed Lines are 95% Confidence Intervals.

It was found that the autocorrelation of pH, SAR, Na, Ca and Mg were significant at 95% confidence intervals with different lag k and hence these were said to be serially dependent. In the other hand, autocorrelation of Cl, NO_3, TDS and Fe were found to be not significant and these were considered as serially independent. Parameter like Cl, NO_3 and TDS are mainly leached to the groundwater through the application of fertilizers and pesticides in the agricultural land. These agrochemicals are usually applied in a particular season of the year. As a result, these parameters have been shown seasonality and now showing serially independence.

Spatial Variation

Five wells, one located in each disrtict of the northwest region, were selected for this analysis. Inspection of means and standard errors of parameter concentrations at five well locations, as presented in Table, reveals spatial variation. Such variation is likely due to small-scale horizontal regional variations and local variations in the aerial application of fertilizer. NO_3 may be denitrified in anaerobic stagnant water, which may cause a large variation spatially. However, parameter concentrations were within the acceptable limits except Fe and Mg. Fe concentration exceeded the permissible limit at all locations while Mg exceeded in flood plain areas of major rivers. Cl, TDS, pH, Na and Mg showed minimum values in upland areas and maximum values at locations near Brahmaputra River, indicates an increasing trend near major rivers. As expected SAR was lower near major rivers because of higher concentration of Ca and Mg at these locations.

Table. Mean and Standard Error of Water Quality Parameters for Five Selected Wells

Well No.	Cl	NO_3	TDS	pH	SAR
4	16.3± 8.6	3.6± 1.9	120.6± 16.4	6.9± 0.7	2.4± 0.5
14	30.5± 5.2	7.2± 2.4	225.8± 24.5	7.2± 0.6	1.3± 0.4
23	23.5± 4.6	0.2± 0.1	181.6± 16.4	7.2± 0.7	1.6± 0.3
27	52.9± 14.7	0.5± 0.2	425.8± 59.4	7.7± 0.8	1.2± 0.2
39	31.6± 6.9	1.2± 0.4	380.6± 32.1	7.7± 0.5	1.5± 0.4
Well No.	**Na**	**Fe**	**Ca**	**Mg**	
4	25.5± 4.5	3.7± 1.1	11.5± 2.3	9.1±3.3	
14	31.3± 4.4	1.9± 0.7	33.4± 9.3	18.5±7.6	
23	37.6± 5.3	2.6± 0.7	27.9± 2.2	18.7±4.9	
27	37.5± 6.7	0.9± 0.3	55.7± 11.7	58.1±16.6	
39	32.6± 4.4	10.9± 3.1	71.3± 12.6	56.7±20.6	

Well 4 located at Thakurgaon, Dinajpur; 14 at Rajshahi Town; 23 at Gakul, Bogra; 27 at Nawabganj, Rajshahi; 39 at Nagarbarighat; Pabna. In this study, groundwater quality variables such as, Cl, NO3, TDS, pH, SAR, Na, Fe, Ca and Mg from northwest region of Bangladesh were analysed statistically. Statistical analyses were performed to check whether the water quality data of concern are normally distributed, exhibits seasonal variation or serially dependent. Spatial variation of parameters was also performed in 5 selected wells located in 5 different districts in the region.

Normality of the data was checked by empirical distribution, skewness test and chi-square goodness-of-fit test and was found nonnormal distribution, rather the data was appeared to be lognormally distributed, which is usually expected for water quality data.

Tests for seasonality were performed using 2– sample tests (Student's t-test and Mann-Whitney U test) and 4-sample tests. It was seen from the results that Cl, NO_3, TDS, SAR and Mg shows seasonal variation because of the variation of precipitation in different seasons.

The parameters like, pH, SAR, Na, Ca and Mg were found to be showing serial correlation. Spatial variability of these parameters was also investigated and found significant variation, especially in case of NO_3, Ca and Mg because of the variations in aerial application of the agrochemicals in different locations.

COMBINED USE OF SURFACE AND GROUND WATER

Since the moment of its construction, the system of water supply to the Moscow City agglomeration has been destined to operate using mainly surface flow. As a result, a network of canals, pipelines, large reservoirs, pumping stations has been constructed, which has changed radically the natural hydrological regime in the basin.

At the same time, intensive water withdrawal by numerous water intake facilities has largely disturbed the ground water flow regime. The Moscow water supply system, which has long been developing intensively, is now on the verge of its environmental exhaustion. Further operation and development of this system necessitates searching for ways to decrease the environmental danger. One of such ways is the combined use of surface and ground waters in a common system of water supply, in which ground water is not used as a permanent source of water, but as a reserve to compensate for water deficiency in dry periods.

WATER RESOURCES OF THE REGION AND THEIR USE

Surface water resources of the Moscow Region, used for its water supply, include the Volga River flow from its source to the Ivankovo Reservoir, the flow of the Volga's right-hand tributary– the Vazuza River, the flow of Moskva River and its tributaries– the Istra, Ruza and Ozerna to the Rublevo Reservoir. The total catchment area is about 56.000 km^2. The total amount of water is 340 m^3/s or 10.7 km^3 on the average.

The alimentation of these rivers is mainly with melt water and summerfall precipitation. The role of groundwater alimentation is not significant. The flow regime is characterised by high spring floods and summer low-flow period. From 65 to 70% of the annual flow passes during the high-flow period. In the Rublevo Reservoir site, near the hydropower station, closing the Moskvoretskaya part of the Moscow water supply system (located on the Moskva River) the maximum mean–annual water discharge of 75.3 m^3/s was registered in 1933–1934, the minimum one (20.8 m^3/s)– in 1921-1922, the average long-term discharge of 544 m^3/s was registered in April 1948–1949, the minimum (monthly) one- 6.6 m^3/s -in August–September of 1938-1939.

At present, the water resource system (WRS) of the Moscow Region includes 8 reservoirs in the neighbouring river basin, as well as river reaches, canals, pumping stations and other hydraulic facilities. The main parameters of the reservoirs, regulating the river flow, are given in Table. The flow regulation ensured the increase in the reliability of water supply in the region. For example, for a 95% reliability (according to the number of uninterrupted years) in the Ivankovo site of the Volga River the regulated flow is 3.0 times as large as the natural one, in the Rublevo site of the Moskva River- 4.0 times, in the Zubtsov site on the Vazuza River- 4.0 times as large as the natural one.

The total safe water yield of the system grew 3.4 times. As it was mentioned above, the surface flow is used as the main source of water supply to Moscow. In the regional water balance, ground waters make only 10% of the surface ones. The main developed aquifers are jointing limestones and dolomites of the Upper, Middle and Lower Carboniferous. Aquifers occur as storeys at depths from 10–30 to 150–250 m and have a thickness of 15–20 to 70–80 m each. The aquifers are interbedded by clayey strata.

In river valleys, hydraulic connection between surface and ground waters is frequently observed. By the beginning of the year of 1988, operational reserves of ground waters of Carboniferous aquifers are estimated at 83.8 m^3/s in the whole Moscow Region.

At the same time, calculations show that the aquifers in the vicinity of the Ivankovo Reservoir allow water withdrawal of up to 8 m^3/s.

This value is hereafter regarded as the maximum level of surface run-off deficit that can be compensated for under the combined management of surface and ground waters.

INCREASED DEMANDS TO THE RELIABILITY OF WATER SUPPLY

The estimation of the available water resources in the region shows that the further development of its water supply system can be implemented either by constructing and interbasinal water transfer, or by involving new ground water masses in the process of water supply, thus increasing the part of the ground water in the existing water supply balance.

One way or another, these measures can have negative environmental consequences, and that is why, they require serious scientific substantiation. At the same time, lately, resource-saving technologies have been introduced all over the world. In solving water supply problems, this is connected with considerable reduction in specific water consumption. Undoubtedly, a similar process will also occur in the Moscow Region.

However, reduction in industrial water consumption will inevitably lead to increased demands to the reliability of uninterrupted water supply, because the less water is spent per unit product, the larger economical damages, caused by water deficit.

This also can be referred to social and economical consequences of interruption in the water supply, since the more economically the water is used for municipal or environmental purposes, the larger the social or ecological damage from the violation of normal water supply.

The reliability of water yield can be increased by two means:

1. Increasing the regulatory volume of the reservoir;
2. Involving additional water sources. The latter can cause new environmental problems. Our investigation have shown, that the risk of interruption in water supply can be decreased by applying a new scientific approach to the combined use of surface and groundwater flow.

METHODS OF THE RESEARCH

The essence of the new approach should not be reduced to opposing surface water to ground water. On the contrary it should envisage the use of more sluggish resources of ground water during interruption in the water supply, caused by river flow fluctuations.

The technique of the analysis of WRS operation, using the resources of both surface and ground waters, developed at the Water Problems Institute of the Russian Academy of Sciences, is as follows.

For every WRS, under its fixed parameters, curves of the probability of water deficit are plotted as a dependence on the adopted firm water yield and its probability. At the same time, for an adopted hydrological series, chronological schedules of the interruptions in water supply are determined. It should be taken into account that the characteristics of interruptions, which appear when river run-off is used for water supply, are to a greater extent dependent on the reservoir management regulations.

The higher the extent of run-off regulation, the better the possibilities of managing water resource deficiency, that is, its depth and duration. The development of WRS management regulations to allow the most efficient utilisation of water resources available in the region is among the most urgent problems in the implementation of combined use of these resources.

The schedules characterise the depth and duration of interruptions. The obtained characteristics are used to calculate the regime of intensive ground water withdrawal during the interruptions for surface flow WRS.

The dependence of ground water level fall on the time is plotted for various values of the firm water yield and its probability. The obtained dependences of ground water level fall during the operation of water intake facilities in variable regime are compared with the dependences, obtained under the constant regime of ground water withdrawal.

From the environmental point of view, the regime of the combined work of two sources of water supply is optimal, when the ground water sources become practically renewable, *i.e.* when the system works with a minimum damage to the environment.

SURFACE AND GROUND WATER FLOW

The technique was verified in the calculations of the combined use of the regulated yield of the Upper Volga part of the Moscow water supply system together with ground water resources of the Dubna area. For the task under consideration, the most interesting are low-flow years and their groups, when there is a deficiency in surface waters.

In order to reveal such years, water inflow to the reservoir, changing from month to month and from year to year, was compared with the firm water yield, assigned in variants. As a result of the calculations, water management indices were obtained, characterising possible water deficit - the volume of unsupplied water in each of the interrupted time intervals (month, year); concrete interrupted years in the investigated hydrological series and the number of deficient months in each of these years were also determined.

Several values of the total firm yield of the Verkhnevolzhskoe and Ivankovo reservoirs were considered in this study. The reliability in terms of the number of uninterrupted years, the duration of the uninterrupted period, and the volume of unsupplied water was determined for each water yield value.

In particular, for the existing firm yield of these reservoirs of 78 m^3/s, the reliability in terms of uninterrupted years amounts to 97%; for water yield of 82 m3/s, the reliability is 95%, and for 86 m^3/s, it equals 91%.

As can be seen from these data, a 10% increase in the firm yield (from 78 to 86 m^3/s) notably reduces the reliability. The number of uninterrupted years in the available hydrological series decreases about threefold.

This is absolutely inadmissible as far as the water supply to a city with a nine-million population is concerned. Moreover, the deficit in the interrupting years abruptly increases and in extremely low–water years can be as high as 20–30 m^3/s, and can rise to 50 m^3/s in some months in the end of the reservoir drawdown. On the other hand, the maximum yield of subsurface water sources in this region that can be used to compensate for the surface water deficiency does not exceed 8 m^3/s. Therefore, we have developed special management regulations of the Upper Volga reservoirs, which allow the deficit of water resources to be reduced at the expense of an increase in the duration of the deficiency period.

Figure 5.6 gives an example of the rules developed for the Ivankovo Reservoir for the combined use of surface and ground water resources for increasing the reliability of water supply to Moscow.

Similar to the dispatcher rules, the working storage of the reservoir was divided into three zones corresponding to three conventional states: wet, dry, and average. Each zone has a so-called line of desirable water reserve in the reservoir and the priority indices for meeting the requirements of different water users.

The reservoir itself, depending on the zone in which its level lies, can be regarded either as a water user (reservoir filling period) or a water source (reservoir drawdown). The patterns of zones, the desired water resources, and the priority indices were determined using a simulation model so as to provide the minimum depth of interruption under low-water conditions. Calculations following the suggested rules showed that it is possible to control the river run-off to provide the specified firm water yield of 86 m^3/s with meeting the condition that the deficit should not exceed 8 m3/s in the years lying within 97% probability.

The occasional use of ground water allowed us to increase the firm water yield of the Upper Volga reservoirs from 78 to 86 m^3/s with the same 97% reliability, thus meeting the existing requirements to the reliability of water supply to Moscow. Owing to the nonuniformity of water withdrawal, the groundwater level can be quickly restored, and ground waters, in fact, become renewable resources.

The developed scheme of the management of combined use of surface and ground waters allow for an appreciable increase in the WRS firm water yield and preserved the required reliability of water supply. Such management is optimum from the environmental viewpoint, since it does not require the construction of new reservoirs, whereas ground water sources act as renewable ones.

It is only natural, that in addition to its ecological expediency, such water management should be evaluated specially for every specific case. The final decision should be made only basing on the results of integral technical and economic evaluation of the situation. This study was supported by the Russian Foundation for Basic Research, proj. no. 00-05-81143.

Table. The Main Parameters of the Reservoirs of WRS For Water Supply of Moscow

Name of the Reservoir	Level Mark		Reservoir Volume		Water Surface	Year of Construction
	Normal Head Level	Dead Volume Level	Total	Useful	Area at Normal Head Level, km^2	
Verkhnevolshsk	206.5	303.0	562.0	487.0	181.0	1830
Ivankovo	124.0	119.5	1120.0	813.0	327.0	1937
Istra	170.0	159.0	183.0	171.5	33.6	1935
Moshaisk	183.0	170.0	235.0	221.4	31.0	1964
Rusa	182.5	169.0	219.8	215.7	32.7	1968
Oserna	182.5	169.0	143.8	140.0	23.1	1968
Vasusa	180.25	170.5	540.0	430.0	106.0	1978
Yausa	215.0	212.0	290.0	130.0	51.0	1978

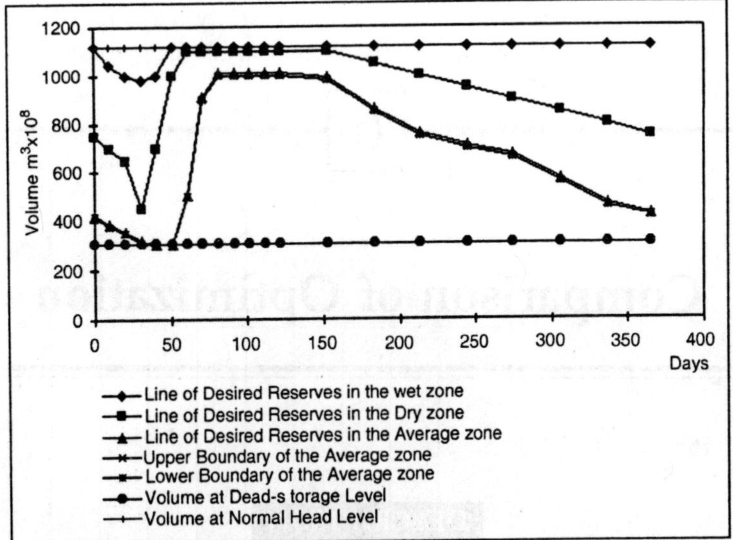

Fig. 5.6 Management Rules For the Lvankovo Reservoir

6

Comparison of Optimization

DESCRIPTION

There are many deterministic models that can simulate long-term runoffs, but most of them have complex structures and requires various observed data for calibration. Therefore, tank models are often preferred in practice since they require only the precipitation, run-off, and evapotranspiration data. The tank model which has a merit of simplicity in data requirement, however, has a drawback, in that it needs much time and effort to calibrate its too many model parameters.

Since the concept of tank models was first proposed, there have been many researches on parameter calibration of the tank models.

Nagai and Kadoya applied three optimization algorithms of Powell's method, DFP method, and QG and proved the excellency of Powell's method and DFP method in parameter identification in 1979. Kim, H. Y. and Park, S. W. evaluated parameter variations with watershed characteristic for six watersheds in Korea.

In 1997, Cooper *et al.* investigated the performance of three optimization algorithms, that is, SA (Simulated Annealing), GA, and SCE (Shuffled Complex Evolution) for calibrating the tank model that contains only two tanks. Ministry of construction and transportation and KOWACO also published many reports about applying tank model to several major dam stations in Korea. Recently, KICT adopted Powell's method for parameter optimization in a tank model.

In this study, three optimization algorithms are tested for the automatic calibration: one nonlinear programming algorithm and two meta-heuristic algorithms (Genetic Algorithm and Harmony Search).

For the proper comparison, sensitivity of optimization parameters of GA and HS that are thought to have great influence on performance of optimization are analysed. And the best values for these optimization parameters are used in comparison.

TANK MODEL

Tank modes are intended for calculation of a run-off with the catchment area of a river substituted by a combination of a number of storage type model vessels, first introduced by Sugawara and Funiyuki.

It is a kind of deterministic, lumped, linear, continuous, and time-invariant model and it can be applied to flood analyses, too. In at least two studies comparing the performances of some of popular conceptual rainfall-run-off models, the tank model demonstrated its capability for modeling the hydrologic responses from a wide range of catchments. Tank model is composed of a few tanks laid vertically in series. The output from the top tank is considered as surface run-off, the output from the second tank as intermediate run-off, the output from the third tank as sub-base run-off and output from the fourth tank as baseflow.

This model structure may be considered to correspond to the zonal structure of the surface and subsurface water. Similarly, the output from the bottom outlet of the top tank could be considered to correspond to infiltration. The outputs from the bottom outlets of the other tanks could be regarded as percolation.

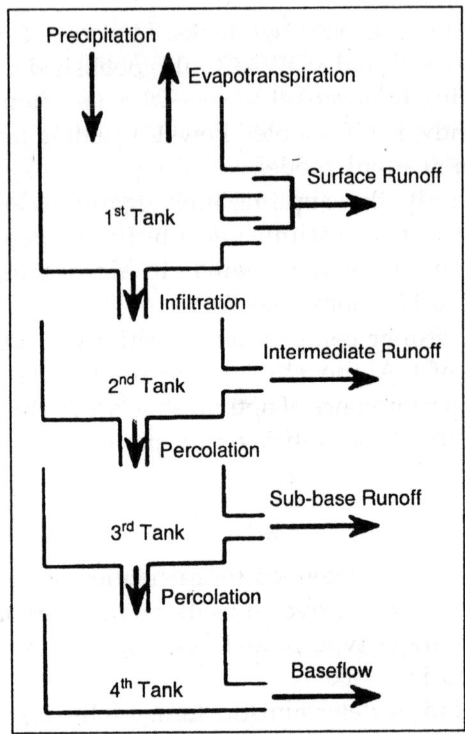

Fig. 6.1 Schematic of Tank Model in This Study.

There are some sorts of variety of the usual tank model. Although many side outlets in first tank can result in representing flood response more precisely, it makes more parameters to be calibrated. In Korea, the common purpose of tank model is to forecast long-term run-off. To build a practical model, a tank model with 4 tanks is introduced. First tank has two side outlets and the other tanks have one side outlet. This structure is the same as the tank model of KICT. Figure 6.1 shows the schematic of the tank model used in this study.

OPTIMIZATION ALGORITHMS

POWELL'S METHOD

Powell's method, the short name of the conjugate direction

method proposed by Powell, is a kind of direct search method and has been applied to many engineering optimization problems for it was known as one of the most efficient method in unconstrained nonlinear optimizations. This technique is a modified or deflected gradient approach. If the function is quadratic in N variables, the procedure converges to the optimal solution in exactly N iterations. The method also gives good results for functions that are more complex than a quadratic form.

Powell's method is much more efficient than steepest descent method. Furthermore, this method chooses successive directions without computation of the function's gradient. This figure reduces the total computational burden drastically. In spite of the above advantages the Powell's method has, it has a typical drawback common in traditional NLP (Non–Linear Programming) algorithms. To get a good optimal solution, many initial starting points must be tried. Unfortunately it is practically impossible to locate the exact starting point for the global optimum.

GENETIC ALGORITHMS

GAs have been developed by John Holland, his colleagues, and his students at the University of Michigan. The primary monograph is Holland's "Adaptation in Natural and Artificial Systems". GAs are search algorithm based on the mechanics of natural selection and natural genetics. By abstracting nature's adaptation algorithm of choice in artificial form we hope to achieve similar breadth of performance. In present times, GAs are known as very efficient heuristic algorithms, surmounting problems of traditional optimization algorithms, and are applied to many engineering problems, inclusive of water resources engineering. In this study, to reinforce the performance of GA, the strategies of multi-point crossover and reserving elite string are added to SGA in this study.

HARMONY SEARCH

HS is a new heuristic algorithm, recently developed by Geem by mimicking the improvisation of music players. Musical performances seek a best state (fantastic harmony) determined by aesthetic estimation, as the optimization algorithms seek a best state (global optimum: minimum cost or maximum benefit or maximum efficiency) determined by objective function value. There are two parameters introduced in HS, that is, HMCR and PAR (Pitch Adjusting Rate). HMCR is introduced to escape from local optima which can occur when all the parts of the global solution doesn't exist initially in HM (Harmony Memory). PAR is adopted for improving solutions and escaping local optima. This option mimics the pitch adjustment of each instrument for tuning the ensemble. As newly developed algorithm, HS has much possibility to be improved. Various experiments are implemented to try to improve the performance of HS in this study.

APPLICATION

STUDY AREA

The study area of this study is Daecheong dam basin, which occupies all but half of the Geum river basin with its catchment area of 4,134km^2. Daecheong multipurpose dam, located 150km above Geum river estuary, 16km north–east from Daejeon, and 16km south from Cheongju, was constructed in December, 1980. Meteorological data including evaporation are collected from 5 stations in Daecheong dam basin, that is, Daejeon, Cheongju, Jeonju, Boeun, and Geochang.

SENSITIVITY ANALYSIS OF OPTIMIZATION PARAMETERS

In the case of GAs or HS, there are optimization parameters; crossover probability, mutation probability, and population size are main optimization parameters in GAs, while HMCR, PAR, and harmony memory size are those in HS. Because these optimization parameters greatly influence the efficiency of optimization algorithms, precedent to comparison of optimization algorithms, sensitivity of these parameters are analysed to find out best values for each optimization parameters for fair comparison. Sensitivity of optimization parameters are evaluated by SSQ (Sum of the Squares). Determined values of optimization parameters, analysed above, are summarized in Table.

Table. Analysed Optimization Parameters

GA	Crossover probability	Mutation probability	Population size
	0.65	0.01	60
HS	**HMCR**	**PAR**	**Harmony memory size**
	0.95	0.3	60

COMPARISON OF OPTIMIZATION ALGORITHMS

Three new strategies of roulette wheel selection, elimination of overlapping harmony, and special training of elite harmony are devised and evaluated to improve the HS's performance. These strategies are evaluated by the criteria of SSQ in parameter calibration of the tank model. On the whole, elimination of overlapping harmony and special training of elite harmony enhance the performance of HS, while roulette wheel selection undermines the performance of HS in experiments.

Two beneficial strategies of elimination of overlapping harmony and special training of elite harmony are simultaneously applied and named MHS. This combined strategy performs better than any other strategy of this study. Relative SSQ of MHS to simple HS is drawn in Figure 6.2. MHS is compared with GA and Powell's method. Initial values of parameters, required in Powell's method, is set as the same as the recommended values by Ministry of construction and transportation et. al. SSQs of GA and MHS are quite smaller than that of Powell's method. To check the properness of calibrated parameters, PEV of three techniques are compared. Though minimizing SSQ is selected as objective function, absolute values of PEV of three algorithms are proportional to the SSQs of these algorithms. Comparison of SSQ and PEV shows the superiority of heuristic algorithms, such as GA and

HS, prior to traditional NLP algorithm (Powell's method) in parameter calibration. This study shows GA or HS can be a reasonable choice for parameter calibration in tank model. Objective function value of GA is a little better than that of MHS after 10,000 iterations. However, objective function value is not the only criteria to evaluate optimization algorithms. The most important goal of optimization is improvement. To answer to the question, how can we get to some good, "satisficing" level of performance quickly, optimization algorithms evolved. Therefore, computation time is another important component to evaluate optimization algorithms. Until 10,000 iterations for GA and 10,000 and 30,000 iterations for HS and MHS, the computational time is recorded with Pentium I I 233Mhz CPU.

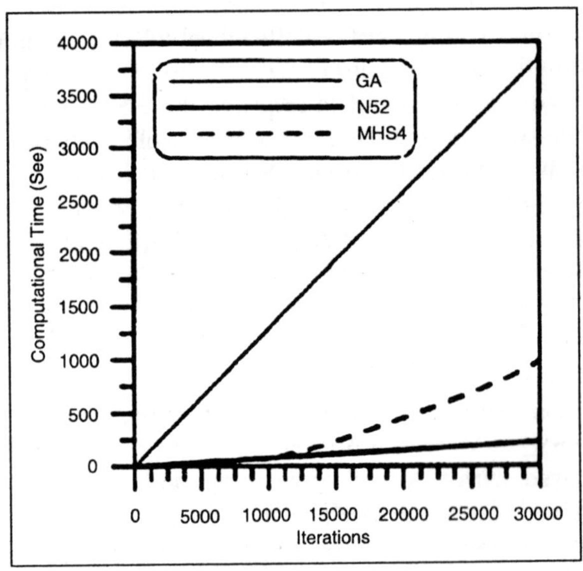

Fig. 6.2 Computational Time to Iterations

Comparison of the computational time shows a relative superiority of HS and MHS. The average CPU time used by MHS is 74 seconds, only 5.75% of GA's computation time of 1,287 seconds. This significantly reduced computational time of MHS compared with GA up to the same iteration number of

10,000 is a big merit. MHS can optimize parameters up to more than 30,000 iterations in less CPU time than GA calculates for 10,000 iterations. The SSQ and PEV of MHS after 30,000 iterations are better than that of GA after 10,000 iterations.

Powell's method, the weakest algorithm in this comparison, has other drawbacks. First, initial values of parameters must be given. To decide initial values is another task for user of this model. Another problem, in fact, most serious problem of Powell's method, is numerical dispersion. In this comparison, only 7 cases are succeeded to give results among 16 cases. These problems are, in a sense, common in traditional NLP algorithms. This handicap of Powell's method supports another reason to select a heuristic algorithm, especially time-efficient MHS proposed in this study.

In this study various optimization algorithms are linked with a tank model to calibrate its 16 parameters. For the GA, crossover probability between 0.5 to 0.8, especially of 0.65, and low mutation probability of 0.01 produce good results whereas population size shows little correlation in performance. This characteristic is the same as previously proved properties in other applications of GAs by many researchers. For the HS, high HMCR, especially from 0.9 to 0.95, contributes excellent outputs, while PAR and harmony memory size show little correlation with improvement in performance. Geem previous sensitivity analysis of HMCR is confirmed in this study.

To enhance HS's performance, three new strategies are devised and attempted in this study. Among these new strategies, two strategies of elimination of overlapping harmony and special training of elite harmony are recommended be used as supplemental algorithms of HS, while roulette wheel selection undermines the performance of HS. Further, the superiority of heuristic algorithms, such as GA or HS, in parameter calibration of tank models is proved by a comparison with Powell's method. Advantages of GA and HS over Powell's method are: less erroneous parameter calibration, needlessness of setting initial values for the parameter values, and independence from numerical dispersions. Although GA gives slightly better results than MHS does in a same number of iteration, MHS performs surprisingly faster than GA in the same computing environment. In a given CPU time, MHS shows better calibration results than GA does for the tank model tested in this study. Up to now, the biggest problem for using tank models in practice has been the parameter calibration. Well-trained engineer's sense has been emphasized in parameter estimation. This study shows a possibility to avoid the problem by the application of heuristic algorithms, especially the Harmony Search algorithm.

AGRICULTURAL WATER SHORTAGE

WATER CRISIS LIMITS THE AGRICULTURE

Since the founding of the People's Republic of China, a large amount of water conservancy infrastructure have been constructed and strongly encouraged the development of the industry and agriculture and urban construction, resisted flood, protected environment, and obviously raised the people's living standards. With the increase in population and rapid development of the society and economy, the water supply obviously cannot meet the water demand and resulted in water shortage. According to statistics, drought disaster bring with more serious affection than flood disaster. Therefore, a critical problem facing by the development of China's society and economy is how to satisfy the water demand of society and economy.

Water shortage resulted in:

- The production from rain-fed land which area is more than 50% of total cultivated area in China has to rely on climate, and is low and unstable. Consequently, the agricultural development has badly been limited.
- The water shortage resulted from water pollution is also remarkable.
- In the northern part of China, a series of the ecological and environmental problems occurred and day and day worsened, including drying up of rivers, bad overdraft of groundwater, decline of groundwater table in broad area, and so on.

- Large amounts of wastewater and sewage without treatment have been directly and indirectly used for irrigation, specially in the northern part of China, and gradually become an important component part of agricultural water use. The use of wastewater and sewage without treatment resulted in the pollution of soil water in farmland and groundwater and the excess content of pollutants in agricultural products which are harmful to the people's health. This kind of harm is very severe.

Except for the water shortage and the water environmental problem, China's water crisis still includes the frequent flood disaster and the low capability of resisting natural disasters. Flood disaster is still a serious danger to the sustainable development of China's agriculture.

WATER CRISIS IN THE 21ST CENTURY IN CHINA

The water demand and supply in the 21st century in China will face:

- Due to too fast increase in water demand and limited increase in water supply, the lag between demand and supply will further be widened;
- Because of the existing water source drying up, water projects aging and inadequate maintenance, and so on, the water supply sharp decline;
- In the northern part of China, the available water supply obviously decrease with deep development of water resources and the water consumption increase in river basin;
- The warmer and warmer climate may aggravate water crisis;
- With industrial and domestic water use increase, water pollution will more and more seriously threaten water supply and safety of water use;
- People don't fully understand the seriousness of water crisis, thereby water projects aren't enough and management is backward. Water supply lag water demand with extravagant water use. The low water use efficiency and water production efficiency further

worsen the water crisis.

The 21st century will be an era of the great development of economy, and an era which China's population will reach the peak of 1.6 billion. Water is an important constraining factor for the economy development. If the problem of water shortage cannot be resolved, the sustainable development of society and economy must be seriously imperiled.

Moreover, the water crisis mainly affect the agriculture development by:

- Competition for water among regions, between industry and agriculture, between urban and rural, will continue for a long time and be more serious.
- Competition for water between agriculture and environment will stand out, specially in the northern part of China.
- With the development of economy and the raise of people's living standards, the water demand by other agriculture, including forestry, animal husbandry, fishery and subsidiary, must extremely increase. Accordingly, competition for water within agriculture sector will be more outstanding.

WATER DEMAND OF NATIONAL ECONOMY DEVELOPMENT

Status of Water Use

The rapid, continuous economy development and improvement of society promote the swift increase of water use. In amount of water use, the toțal water use increased by about one fourth, from 443.7 B m^3 in 1980 to 559.1 B m^3 in 1999. But there were different growth rates in different sectors. The percentage agricultural water use in total water use decreased from 83.4% to 69.2%, while the percentage of industrial water use increased from 10.3% to 20.7% and the percentage of domestic water use increased from 6.3% to 10.1%. This change indicated the structure of water use was profoundly transforming. The statistics of water use in 1980 and 1999 are shown in Table.

Table. Water Use in 1980 and 1999 in China

Year	Item	Industry	Agriculture			Domesticity			Total
			Irrigation	Other Agriculture	Subtotal	Rural	Urban	Subtotal	
1980	Amount (B m^3)	45.7	358.1	11.9	369.9	21.3	6.8	28.0	443.7
	Percentage (%)	10.3	80.7	2.7	83.4	4.8	1.5	6.3	100
1999	Amount (B m^3)	115.9	356.4	30.5	386.9	29.6	26.7	56.3	559.1
	Percentage (%)	20.7	63.7	5.5	69.2	5.3	4.8	10.1	100
Increment (B m^3)		70.2	-1.7	18.6	17.0	8.3	19.9	28.3	115.4

Note: Water use by theromal power plants included in industrial water use; water use by commercial vegetables included in urban domestic water use; water use by domestic livestock included in rural domestic water use.

THE DEVELOPMENT OF IRRIGATION AREA AND AGRICULTURAL WATER USE

Rural economy rapidly developed and grain production steady increased, but the increment of agricultural water use was limited. Though the actual irrigation water use is affected by many factors including yearly hydrologic and meteorologic conditions, the general trend of water use in the whole country still can be concluded.

The decrease of irrigation water use was mainly due to effective agricultural water saving and shortage of water supply for agriculture in water scarcity regions. With the development and popularization of agricultural water saving, the efficiency of agricultural water use is being raised. The actual irrigation area increased by about 8,200 thousand hm^2 from 1980 to 1999, but the change of irrigation water use was tiny. In general, the average water use per hm^2 decreased from 8,750 m^3 in 1980 to 7,270 m^3 in 1999. The water use of per unit area decreased by 17%. These agricultural statistics are shown in Table. The water use by forest, herd and fishery fast increased while irrigation water use decreasing. Sequentially, the water demand by ecological environment had better been met.

Table. Irrigation Area and Water Use

Year	Irrigation Water Use (B m^3)	Actual Irrigation Area (10^3 hm^2)	Unit water use (m^3/hm^2)
1980	358.1	40 920	8 750
1999	356.4	49 090	7 260
Increment	-1.7	8 170	-1 490

WATER DEMAND

Based on the analysis of the relationship between water resources and the sustainable development of national economy, the characteristic of water resources, factors of water resources rational allocation supporting the sustainable development, status of water use and supply, as well as the existing problems, the water demand before the mid-21st century have been predicted. The prediction on water demand by industry, agriculture and domesticity in different target years are shown in Table.

Table. Prediction on Water Demand (B m^3)

Year	Urban	Rural	Industry	Irrigation	Other Agriculture	Total
2010	40.5	30.2	149.8	387.9	34.0	642.4
2030	64.1	30.9	191.1	387.2	38.5	711.8
2050	81.5	30.6	199.8	377.5	42.5	731.9

Table indicates the basic increasing trend of water use: faster increasing of industrial and domestic water use, little increasing of agricultural water use, steady growth of proportion of industrial and domestic water use in total water use, decrease of proportion of agricultural water use. However, because of natural condition, the agricultural development in China has to rely on irrigation. So the proportion of irrigation water use in total national economy water use always is highest among the sectors.

According to the status and the development trend, three main problems, such as flood disaster, water shortage and water environment worsen, especially the water shortage will more and more heavily limit the development of agriculture, society and economy. How to resolve the contradiction between the sustainable development of agriculture and water supply is one of the focus related to the sustainable development of society and economy.

AGRICULTURAL WATER CRISIS

The agriculture development in China relies more and more heavily on water sources. Based on the analysis of the existing problems and the improvement of society, science and technology, rational development and high efficient use of agricultural water resources is unique strategy accounting for water crisis involving in grain supply to 1.6 billion Chinese in the mid-21st century. Water crisis is affected by many factors. The following countermeasures should be taken:

Energetically Increasing Input to the Construction

While the development of water saving in agriculture, the input to the construction of water conservancy infrastructure should be increased so as to develop various water source for irrigation. Some important water projects for the development of national economy should be constructed to change the status of water shortage in the north part of China and the coastal regions, to improve the production condition of the important production bases of grain and cotton, and to promote the agriculture development.

Developing rain irrigation is very important not only to the agriculture development on rain-fed land, but also to the high efficient use of irrigation water and the relaxation of contradiction between the demand and the supply. The utilization degree of rain and flood during flood season should be raised by structure and non-structure measures in order to combine flood control with generating benefit, *i.e.* utilizing resources while controlling disaster. It will play an outstanding role in relaxation of water crisis and flood disaster. The treatment and reuse of wastewater and low quality water will not only relax the water crisis in irrigation, but also improve and protect the ecological environment.

Vigorously Developing High Efficient Use of Agricultural Water Use

At present, the production efficiency of irrigation water is still less than 40%, accordingly which decrease the utilization efficiency of rain and irrigation water and worsen the water crisis. The water production efficiency can be improved by taking synthesis agricultural measures to decrease crop's invalid water consumption. These agricultural measures include: planting the crops and breed with the characteristics of low water consumption and high production; improving utilization ratio of soil water and protecting soil moisture; high efficient allocation and control of water and fertilizer; covering planting; improving crop's drought resistance and water utilization efficiency by chemical measures.

The development of high efficient use of agricultural water use relies not only on technology, but also on the following sufficient studies: studying and establishing the strategy and target of the development of agricultural water saving; study on the basic gist putting forward the general target and task of agricultural water saving; for different stage and region, study on the requirement on agricultural water saving by national economy development, the national economic base supporting the development of agricultural water saving, farmer's income, investment ability and interesting, as well as the development strategy and technologic principle; for different stage, study on the mechanism prompting agricultural water saving and its change and control measures; study on the stage of industrialization of agricultural water saving.

The development of high efficient use of agricultural water use must be sustainable and accord with the following three criterions: firstly, sustainablity–can effectively protect, rationally and high efficiently utilize water resources and make the utilization sustainable; Secondly, efficiency-can markedly raise economic efficiency, including water saving, production increase, land saving, energy saving, ecological environmental benefit, profit of non-agricultural capital input, and so on; Finally, high scientific and technological characteristic-should be scientific and technologic and can fully meet the requirement on high efficient use of agricultural water use at different development stage.

High Efficient use of agricultural water use is the foundation supporting the sustainable development of agricultural and the sustainable utilization of water resources.

Energetically Improving and Protecting

The simple improvement with protection target should be transferred to the improvement with exploitation target in order to combine tightly ecological benefit with economic benefit. The exploitation should be combined with and promotes the improvement. Strengthening the construction of water conservancy infrastructure on farmland in order to raise the farmland's ability of resisting water-logging and saline disasters, and rationally utilize the water resources. The pollution of the agricultural water resources should be effectively prevented and improved by the rational and effective use of fertilizer and bad water source.

Intensifying the Management For High Efficient

It is recognized in the World that 50% of potential of irrigation water saving is from the improvement of management. There is a very big gap in management plane of water resources between China and the developed countries. Firstly, the policies and codes prompting high efficient use of water resources are not perfect. There is short of mechanism and management measures encouraging high efficient use of agricultural water use. Secondly, the technical system supporting high efficient use of water resources has not been established.

The potential of water resources far has not been fully developed. Thirdly, there is short of engineering measure which can support the improvement of management, because of the low project standards of many agricultural water projects.

Therefore, to resolve the water crisis, the water management must be changed from an extensive pattern to an intensive pattern, including:

- Establishing the management system supporting rational and high efficient use of water resources to make the unified management of water resources available.
- Establishing the price system prompting rational use of water resources to form a sound market mechanism for the management of water use.
- Establishing the guarantee system prompting high efficient use of agricultural water use, including investment mechanism.

Strengthening Scientific Study and Research

With the contradiction between the demand and the supply is day and day outstanding and the national economy is developing, many technical problems related to the high efficient water use are waiting for resolving and new technical problems appear in succession. All the problems need to be profoundly studied. The technical integration and innovate system supporting the high efficient use of agricultural water use should be established. The science and education should benefit the development of water resources. The development of series products of high efficient water use should be strengthened. The efficiency of water use should be raised.

Reforming the Mechanism of Water Management

The technical reform and management system reform of irrigation districts must be pressed on to reach the target of high efficient use of water resources and solidify the agricultural production on irrigated area on which the crop's production account for third-fourth of the entire country. The technical reform of irrigation districts is basis of high efficient water use, and the management system reform of irrigation districts is its assurance. The management system reform of irrigation districts will benefit not only for the exertion and increase of value of state's capital, but also for the sustainable development of irrigation agriculture and high efficient water use, even for food security. So it is one of the important measures accounting for the water crisis, and is also a key component of the prosecution and management system reform in agriculture and rural. This reform should be in line with the development of market agriculture and integration of prosecution to be one part of the socialism market economy.

Because of the limit of natural condition, immense requirement of the society and economy development, weak economic base, and so on, the sustainable development of agriculture is facing severe water crisis. With the population increase, people's living standard raise and cultivated land decrease, the press on cultivated land, specially on irrigated cultivated land, will be more and more heavy. Water will dominate the fortune of China's agriculture and the survival and development of 1.6 billion Chinese in the 21st century.

Based on the analysis of water crisis, social practice and the advancement of science and technology, the essential measure accounting for the water crisis faced by the sustainable development of agriculture is high efficient use of agricultural water use. The raise of water use efficiency and water production efficiency is key strategic measure related to the survival and development of 1.6 billion Chinese in the 21st century. The development of agriculture and water is a huge system engineering. The water crisis must be roundly resolved from system viewpoint. It should be realised the water's complexity. Don't keep eyes only on water. The problem of water resources should be roundly resolved from system viewpoint and by means of carrying out the comprehensive prevention and cure strategy, to reach effect which is more than the addition of single strategy's effect. The utilization efficiency and production efficiency of water resources must be raised by means of system engineering and carrying out the comprehensive measures including water projects, agricultural biologic measure, modern management technology, information engineering technology and meteorology, to basically resolve the contradiction between water supply and demand, and to support the rapid development of society and economy in China in the 21st century.

CURVE NUMBERS, RECENT DEVELOPMENTS

The Curve Number procedure of the U.S. Dept. of Agriculture, Natural Resources Conservation Service (NRCS) (formerly Soil Conservation Service, SCS) has elicited questions and concern since its conception. This arises, for the most part, because users read into the procedure what they wish was covered. The actual intent of the procedure is often disregarded. Too, the basic reference for Curve Numbers, the National Engineering Handbook of the SCS, has been revised several times, not always by individuals or committees that understood the significance of their statements. The Natural Resources Conservation Service and the Agricultural Research Service, both agencies of the U.S. Department of Agriculture, formed a joint work group to assess the state of the Curve Number procedure and to chart its future development.

The joint work group recognized three distinctly different modes of application for Curve Numbers:

1. Determination of run-off volume of a given return period, given total event rainfall for that return period;
2. Determine direct run-off for individual events, explaining the variability from event to event, as used in continuous simulation models;
3. Determine infiltration rates for short time intervals as used with unit hydrograph development of flood hydrographs.

The first mode of application represents the historical basis of the procedure, so receives the most attention. Use as a surrogate for an infiltration is very common and follows from the historical basis, so must be considered. The application in continuous simulation models is an extension beyond the scope of the committee. Discussions within the committee made it apparent that a portion of the difficulty surrounding the procedure was attributable to the presentation of the procedure in the National Engineering Handbook. The first task then was to rewrite those portions of the Handbook pertaining to the procedure. Problems identified ranged from incorrect and misleading statements to incomplete documentation. For example, it was incorrectly stated that S includes Ia, whereas it can be shown mathematically that S does not include Ia.

Fortunately, this is only significant for continuous simulation. Another example was a table that related antecedent rainfall to antecedent moisture condition (AMC). This was not intended to have nationwide application, though it was treated as such. Folklore concerning Curve Numbers could also be attributed to problems with documentation. A folklore example is that the Curve Number Run-off Equation is an infiltration equation.

In rewriting the Curve Number portions of the Handbook, the work group agreed that:

- Committee must Abelieve in≅ concepts expressed;
- References will be included if possible; and
- Results must be technically defensible.

The results of the rewrite include such items as:

- Reference to Antecedent Moisture Condition (AMC) was removed. Variability is incorporated by considering the curve number as a random variable and the AMC–I and AMC–III conditions as bounds on the distribution.
- Reiteration of desirability of locally determined curve numbers. This was part of the original documentation but tended to be neglected.
- Explicit expression of Curve Number run-off equation as a transformation of rainfall frequency distribution to run-off frequency distribution. This was demonstrated in the original documentation but again was often neglected.
- Expression of AMC-I and AMC–III as measures of dispersion about the central tendency (AMC II). This is a corollary of treating the CN as a random variable.
- Mathematical proof showing that S does not include Ia. This is only significant because of the previous missunderstanding.

As the work group progressed on the rewriting, they reached a level of agreement on principles allowing work to begin on two other areas of need. These were to reconsider the hydrologic soils classifications recognizing the vastly expanded data base available today and the capabilities of modern computers, and to reconsider the tables of curve numbers in terms of the expanded rainfall-run-off database available.

HYDROLOGIC SOIL GROUPS

There has been a vast increase in basic soils property data since Musgrave first proposed the concept of hydrologic soil groups in Handbook of Agriculture. The data are now available in an electronic database. Modern tools of data mining were explored for analysis of this mass of data. Both neural networks and fuzzy sets were tried with fuzzy sets being adopted.

Originally, Soil Hydrologic Groups were assigned to soil series and phase of series by soil scientists based upon their interpretation of the published criteria. The soil scientist=s interpretation of the published criteria has varied across time and between states or regions. Thus, the hydrologic group criteria are not applied consistently across the United States. This is most evident in the comparison of soils with similar soil hydrologic and physical properties and dissimilar hydrologic group placement.

The Hydrologic Soil Groups are A, B, C, D and dual groups A/D, B/D and C/D. Soils in hydrologic group A have low run-off potential. Soils that have a moderate rate of infiltration when thoroughly wet are in hydrologic group B. Hydrologic group C soils that have a slow rate of infiltration rate when thoroughly wet. Soils in hydrologic group D have a high run-off potential. Dual Hydrologic Soil Groups (A/D, B/D, and C/D) are given for certain wet soils that could be adequately drained. The first letter applies to the drained and the second to the undrained condition. Soils are assigned to dual groups if the shallow depth to a permanent water table is the sole criteria for assigning a soil to hydrologic group D.

A model or rule based automated system that provides for objective placement of soils into Hydrologic Soil Groups was developed. The fuzzy system model for assigning soils to hydrologic soil groups is based on the published hydrologic group assumptions and criteria. The soil surface is taken to be bare and the soil is not permanently frozen. The soil physical and hydrologic characteristic which make up the hydrologic grouping criteria are the depth to permanent water, depth to a restrictive layer, minimum saturated hydraulic conductivity in the soil=s upper 100 cm, and the soil=s texture.

There are three components to the fuzzy systems model: the Property, the Evaluation, and the Rule. The Property is an

SQL (Standard Query Language) statement that retrieves the needed soil data from the soil survey database. An example of a Property is the depth to a restrictive layer.

The Evaluation=s function is to apply the data received from the SQL statement to a statement of the property=s relevance to the soil=s hydrologic grouping. In the case of the depth to a restrictive layer, the Evaluation determines the fit or truthfulness of the statement, AThe run-off characteristics of the soil increases as a soil=s depth a restrictive layer becomes shallower. At some depth, the restrictive layer in the soil has a maximum contribution to run-off and the Evaluation is true. The result of an Evaluation is some number between 0 and 1. This number represents the truthfulness of the statement being evaluated.

The closer the number is to 1 the closer the soil=s property fits the grouping criterion. Conversely, the closer the number is to 0 the less the soil property=s contribution to the hydrologic grouping of soils. In the restrictive layer example, an Evaluation output of 1 would mean that the soil=s restrictive layer is shallower than 50cm. Any output less than 1 would mean that the depth to any soil restrictive layer is greater than 50cm. This numeric output from the evaluation is passed to the Rule. The Rule is the third component of the fuzzy system model. The Rule serves two functions that result in a soil=s Hydrologic Soil Group placement. The first function is to provide tools for the construction and implementation of the grouping system=s model and to bring the various hydrologic grouping criterion evaluations together into a single Hydrologic Soil Group model. The second is to convert the model=s numeric output into a Hydrologic Soil Group.

The model was applied to 1828 unique soil phases using data from Kansas, South Dakota, Missouri, Iowa, Wyoming, and Colorado and the correlation between these soils= assigned and modeled hydrologic grouping was analysed. Table shows a detailed comparison by Hydrologic Soil Group between the currently assigned HSG and the modeled HSG. The correlation between the assigned and modeled HSG A and HSG D soils is higher than the correlation between the assigned and modeled HSG B and HSG C soils.

There are several reasons for the poorer correlation between the assigned and modeled groups B and C. The first is that of the boundary condition which occurs when a soil has properties that do not fit entirely into a single hydrologic group. In this case, the soil scientist may have placed the soil into one HSG while the model placed the soil into an adjacent group. Groups B and C are the most prone to this error because they are bounded by two groups whereas HSG A and D are only bounded by one group. Another source of correlation inconsistency is that the assigned HSG may be relatively correct, but the data in the database may not support the corresponding HSG determination by the model. Finally, correlation inconsistencies can be attributed to the fuzzy modeling of the subjective Hydrologic Soil Group criteria.

Table. Correlation Frequency Between Assigned and Fuzzy Modeled Hydrologic Soil Groups

Current	Number	Fuzzy hsg assignment frequency						
HSG	Of Soils	A	B	C	D	A/D	B/D	C/D
A	155	0.9	0.08	0	0.01	0.01	0	0
B	821	0.25	0.54	0.17	0.02	0.01	0	0
C	405	0.04	0.25	0.34	0.31	0	0.03	0.04
D	404	0.02	0.05	0.05	0.64	0.06	0.1	0.08
A/D	1	0	0	0	0	0	0.55	0
B/D	29	0.1	0.07	0.07	0	0.1	0.31	0.1
C/D	13	0	0.08	0.08	0.39	0		0.15

CURVE NUMBERS FROM RAINFALL-RUN-OFF DATA

The watershed research programme of the USDA, Agricultural Research Service is a continuation of research initiated by the Soil Conservation Service. Much of the data collected should be directly applicable to the determination of curve numbers and to explaining the variation of curve number with the soil-cover complex. In addition, Prof. R. H. Hawkins of the University of Arizona had been developing the world=s largest event rainfall–run-off data base with software to analyse the data. Prof. Hawkins was added to the ARS/NRCS work group. The ideal method of determining curve numbers from observed data is elusive due to the stochastic nature of the variable.

Our decision to emphasize the concept that the run-off equation serves to transform a rainfall frequency distribution into a run-off frequency distribution led to use of frequency matching. That is, curve numbers were determined by use of rainfall of a given return period with run-off of the same return period. These data may or may not come from the same storm. That is, frequency transformation leads to treating ordered pairs. In his analysis, Hawkins recognized that not all data sets are adequate to define a curve number and some watersheds do not even perform according to the Curve Number run-off equation.

He developed a graphical procedure in which the calculated curve number is plotted versus the precipitation used in calculating that curve number. In part, this plot is in recognition that, due to the random nature of the curve number, for a watershed with a given Atrue@ curve number, the actual event curve number will range above and below that Atrue@ value. For small rainfall events the event Ia will vary above and below the Atrue@ Ia. Curve numbers can only be determined if there is run-off, so if the event Ia is low, run-off will occur and a CN computed. If the Ia is high, no run-off occurs so no CN can be computed.

Thus, the process of computing CN for small events biases the CN towards high values (low Ia). The CN vs. P plot displays this bias and the storm magnitude at which the bias becomes insignificant. The concept of the CN method being a transformation between a rainfall-depth distribution and a run-off depth distribution is applied in treating rainfall and run-off data. The rainfall depths and the run-off depths are sorted separately and then re–aligned on a rank order basis to form

P:Q pairs of equal return period. The individual runoffs are not necessarily associated with the original causative rainfalls.

When CNs are calculated from real storm data as outlined above, a secondary relationship almost always emerges between CN and storm rainfall depth itself. In most of these cases, these calculated CNs approach a constant value with increasing rainfall.Three variations on this theme have been observed, however, and are described in the following:

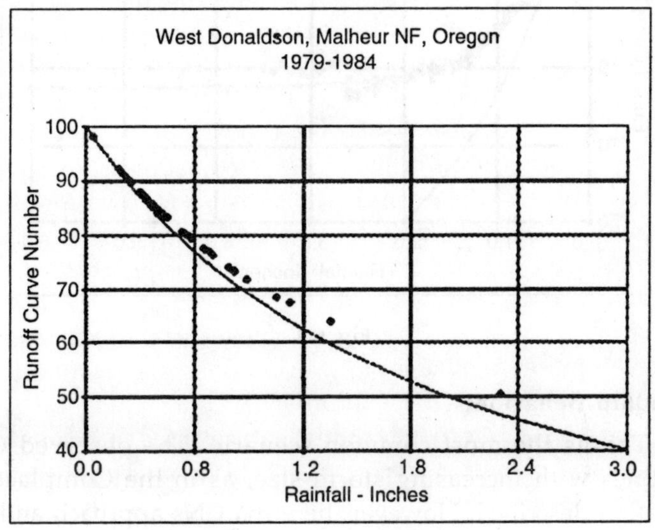

Fig. 6.3 Complacent Behaviour

Complacent Behaviour

Here the observed CN declines steadily with increasing rainfall depth, and with no appreciable tendency to achieve a stable value. An example of this is given in Figure 6.3. Curve Numbers cannot be safely determined from data which exhibit this pattern, because no constant value is clearly approached. This Curve Number behaviour has been found to indicate a partial source area situation where the source area fraction may be quite small. In these cases the run-off is more properly modeled by the linear form Q=CP rather than by the Curve Number run-off equation.

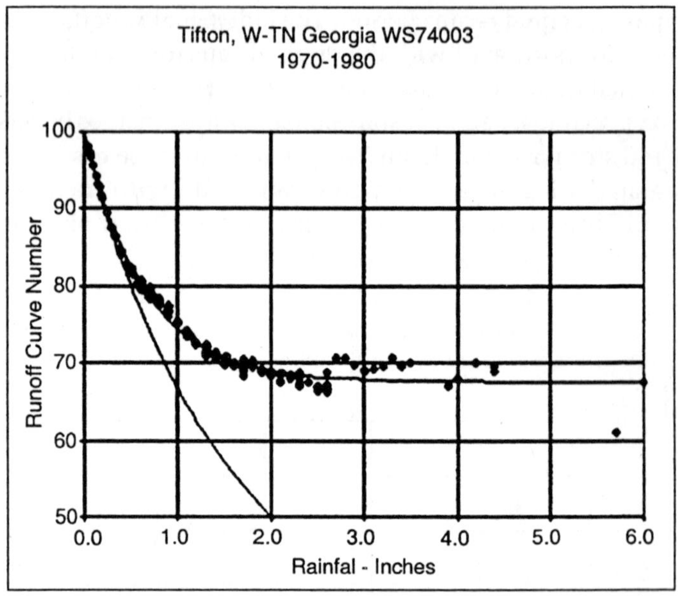

Fig. 6.4

Standard Behaviour

This is the most common scenario. The observed CN declines with increasing storm size, as in the Complacent situation described. However, here the CNs approach and/or maintain a near–constant value with increasingly larger storms. The run-off itself may arise from a variety of source processes, including overland flow and rapid subsurface flow. An example of this pattern is given in Figure.

Violent Behaviour

The distinguishing feature here is that the observed CNs rise suddenly and asymptotically approach an apparent constant value. There is often accompanying Complacent behaviour at lower rainfalls. From a source process standpoint, this could be a threshold phenomenon at some critical rainfall depth value. An illustration of this is given in Figure 6.5. Rietz and Hawkins used their large electronic database of rainfall-run-off data to determine data-defined CNs calculated from local rainfall-run-off data. Their study attempted to develop a better understanding of a watershed=s land use as manifested in its CN. The land use variable was isolated in a large data set of small watersheds, and land use CN=s were calculated and analysed for each of these watersheds at a local, regional, and national scale.

Data in this study contained detailed land use information on 177 watersheds covering 2,455 years of record and 32,891 events. Watershed land uses analysed in this study were: alfalfa (closec–seed legume), corn (row crop), cotton (row crop), desert shrub, fallow, forest, grassland, meadow, oats (small grain), pasture, range, sage brush, sorghum (row crop), soy bean (row crop) and wheat (small grain). Many of the cultivated watersheds had a different land use year to year due to crop rotation. When one watershed had several land uses throughout the period of record, data were segregated by date of land use and CN calculated for each land use. Curve Numbers for each land use on each watershed were determined using the Asymptotic data-derived procedure with ordered P:Q pairs.

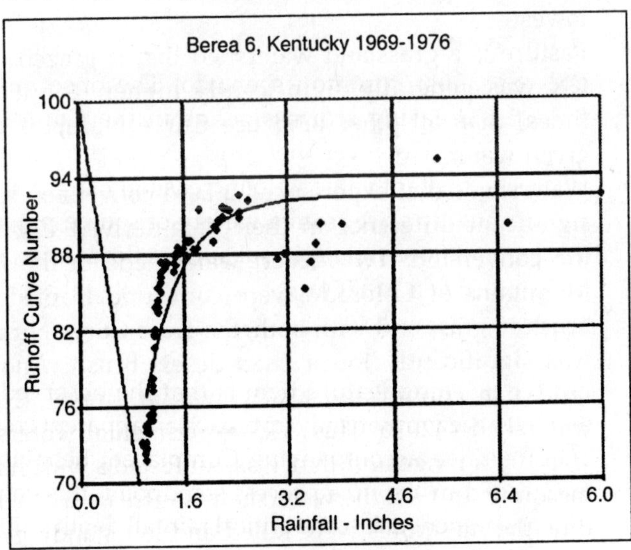

Fig. 6.5 Violent Behaviour

Differences in land use CN at the local level can be attributed solely to a change in hydrological response of a watershed due to land use because all other input variables of the Curve Number model are constant. Soil type, climate, and morphology are fixed, and watershed land conditions are constant for that land use.

Of the 177 watersheds used in this study, 53 were found to have more than one land use during their period of record. When ANOVA independent group tests were performed 96.23% of the watersheds (51 out of 53) were found to have significant differences among their land use Curve Numbers.

The results were:

- Meadow Curve Numbers were almost always the lowest CN for a watershed. Of the 19 watersheds that had a meadow in their crop rotation, 13 had meadow Curve Numbers significantly lower than any other land use CN within that same watershed. Three others had meadow Curve Numbers that ranked second lowest.
- Pasture is a grassland watershed that is grazed. This CN was generally significantly higher (6 out of 8 times) than all other land use Curve Numbers on a given watershed.
- Watersheds that experienced a land conversion had a significant difference in their data-derived CN after the conversion. Two desert watersheds in the Boco Mountains of Colorado were converted from desert brush to grass and with both, the grass Curve Number was significantly lower than desert brush. Another land use conversion study was at Riesel, Texas watershed #42036. This ARS experimental watershed was 100% rangeland that was Ainfested≡ with honey mesquite. In 1972 the watershed was treated, at which time the mesquites were killed and left standing. The Curve Number following mesquite killing was statistically higher than for live mesquites.

In order to compare land use Curve Numbers across all locations, a land use had to have been applied in more than one location, which was found to be the case with 11 single land uses. Seven of these 11 land uses (63.6%) exhibited a significant difference. The rank order of Curve Numbers was: forest and meadow were among the lowest average Curve Number, row crops or small grain were in mid-range, and desert brush exhibited the highest average Curve Number. Small grains (wheat and oats) were grouped together as were row crops (soybean, corn and sorghum). It was surprising however, that the average Curve Number for range was the third lowest average Curve Number and that row crops had a lower average CN than small grains.

The results of the work group are:

- The basic description of the Curve Number procedure is more clear, more consistent, and is in a more technically defensible state.
- The Hydrologic Soil Groups are associated with soil physical properties through fuzzy set procedures.
- Procedures for determining Curve Numbers from local data and interpretation of the results are much better established.

However, more work is needed. For example, we still need to work at verifying, adjusting, and correcting the table of CN. Regional variation in Curve Numbers should be explored. Finally, the whole issue of the application of Curve Numbers in continuous simulation models (CREAMS, GLEAMS, EPIC, SWAT) may be quite different from the design storm Curve Numbers of NEH–4. If there is a relation, that relation should be determined.

7

Ensure Sustainable Development of Water Resources

INTRODUCTION

In decades, the emphasis of water resources management was put on water supply management by ways of searching and developing new water sources, water transfer and distribution, and water treatment and so on. With global shortage of fresh water resources, deterioration of water quality in various water bodies, increasing of cost for developing new water source, as well as population growth and water demand growth, etc., the shortcomings of supply–oriented water management were been aware. Providing water free of charges or heavily subsidized in the past have already led to serious mis–allocations, inefficient using and overexploitation of water resources.

In recent decades, with rapid economic growth in Changjiang Valley, the water demand from all sectors increased quickly. Statistic show that the total amount of water use increase from 135 billion m3 in 1980 to 164 billion m^3 in 1993, while the amount of industrial water using was doubled. Water using increasing gave more pressures to water supply projects in Changjiang River Valley, and also led to water pollution problems, especially in the Changjiang Delta region.

Due to the fact that charges for water supply are generally lower than the cost or simply free of charge, water supply units have been in a state of fund deficit. This has led to inadequate maintenance of water supply system, therefore resulting to inefficient water using. Those issues have been already recongernized by the water resources managers in Changjiang River Basin. Efforts are made to execute water demand management, which is considered crucial for sustainable development of water resources in Changjiang River Basin.

CONCEPT OF WATER DEMAND MANAGEMENT

Definitions

Demand management is defined as the development and implementation of strategies to influence demand for the efficient and sustainable use of the scarce water resources. The aims of demand management are to safeguard the rights of access to water for future generations, to limit water demands, to ensure equitable water distribution, to protect the environment, to maximize the socio-economic output of one unit water, and hence to increase the efficiency of water use and so on.

Tools and Measures

The tools and measures for demand management could be on economic, legal, administrative, educational and technical aspects, etc. From legal angle, the demand management is to formulating and implementing water law and regulations, while economic measures used in the demand management is to charge water using and apply different price policy. From the technical view, those technical measures, such as: drip irrigation, leakage control, canal lining etc., can be considered as water demand management also.

CURRENT SITUATIONS FOR WATER DEMAND MANAGEMENT

Legal Aspect

In order to ensure water resources equitable distribution and allocation, to improve water use efficiency and to make up the cost in the operating, maintaining and renewing the water supply projects, the regulation of water pricing and water fee using for water supply projects was issued by the State Council in 1985, and it indicated that there is no more water supplied for free.

In this regulation, it is stipulated that water users, whom extract water from water supply projects, such as reservoirs, cannels, etc., should be charged by the managing companies, regardless the purposes of water use. The principle for drawing water tariffs as well as the ways of collecting water fees and using the money are also ruled in this regulation.

In 1988, Water Law was published, it stipulated officially that water resources is also a kind of economic goods, those who take the water from lakes and rivers or ground aquifers, should get water usage licenses and be charged in some degree, so called water resources fee.

In 1993, the State Council draw up the regulation of implementing procedure for water usage licenses, it is made clear that each unit or individual who take water directly from lakes, rivers or ground aquifers by constructing projects or pumping facilities etc., should apply water usage licenses, and be charged based on the amount of water taken, which be contracted in the license.

Anymore, the regulation of annual examining water usage was set up, it rules that the water administrative departments at all levels should examine all water users annually within its authority, in terms of the amount of water taken, the water quality of discharge, water saving measures taken as well as the amount of water applied for the next year. Anymore, the water administrative departments have the rights to examine all index mentioned above at anytime within the year, and if the amount of water taken in the year exceed that contracted in the license, it should be charged at higher rate, also several penalties are set up in case of no compliance with the regulation.

Economic Instruments

The current economic instruments which are being put into practice in Changjiang River Basin have three: water resources fee, water fee charged by water supply projects, and tape water charge.

Water Resources Fee

Except for Shanghai City and Hunan Province, other provinces in Changjiang River Basin have already set up the regulations of collecting water resources fee based on the local situation, all water users have been registered and paid for water taking. Those who construct the new projects or expand the scale for water taking should go to corresponding water administrative departments for fore-apply and applying for the licenses.

The rates of water resources fee vary with the provinces, and being set up by provincial water administrative departments. The standard for water resources fee setting are mainly dependent on the types of water use and water supply sources. At present, the rate of water resources fee is set too low (for surface water taking, it cost 0.02-0.05 RMB for 1 m^3, while it cost 0.03-0.05 RMB for 1 m^3 groundwater drawing), and not high enough to show the functions of economic instrument.

Especially the rate between surface water taking and ground water drawing are too close, and led to ground water being over–developed in the regions whose surface water are highly polluted.

For instants, in Changjiang delta plain, land subsidence have already occurred in some area, and threatened the safety of the highway from Shanghai to Nanjing. The maximiue depth of land subsidence even reached 1 m in some region. The reasons causing the problem are that the groundwater was over-taken by the companies and enterprises due to the surface water highly polluted and eutrophication. The cost of taking, treating surface water is higher than that of ground water extraction (including paying higher water resources fee).

To avoid the worse state occurring, Jiangsu provincial water administrative department adjusted the strategies for ground water management, and the water resources fee for ground water will be raised in a short period, and the applying procedure for digging new wells will be more stricter, in addition, the plan for reducing the amount of groundwater extracting annually have already been made and been implemented.

It has proved that these measures have effects on the control of land subsidence. Stricter regulation for ground water extracting in those regions will be draw up soon.

Water Fee Charged by Water Supply Projects

It has been made clear in the Water Law that those who use water provided by water-supply projects shall pay water fee to the supplying unit. The activities charged include using facilities to take water from reservoirs, irrigation canals, manmade channels, etc. for different water use purposes.

At present, water fee is considered as a kind of administrative charge, and each year the government gave subsidies to those projects. To lighten the burden of farmers, some local governments even cancel the water fee for irrigation purpose. The standards for collecting water fee are also set far below the cost, and the non-block rate is commonly practiced. Therefore it is no good for encouraging efficient water use, and water supply units also can not receive sufficient payment for their services, consequently, they are not able to maintain their system adequately, hence led to system at the edge of break-down and inefficient water use.

Another problem occurring in the implementing procedure of this regulation is that agreement between different administrative regions is hard to be reached in case of that the water supply project and water users belong to different administrative regions.

For example, the Danjiangkou project on Hanjiang River, which is in charge of by Changjiang Water Resource Commission, MWR, is a multiple purpose project. It has several functions such as: flood control, power generation, irrigation and water supply for industry.

More than 30 enterprises and tape-water supply companies from Hubei Province and Henan Province are drawing water from the reservoir without any payment, due to the agreement can not be reached between local government and the project managing body, until now (20 years after the regulation drawn), the affair is still under discussion.

Water Price for Tape Water

Tape-water supplying industry is managed by government in China, and the charge for tape-water is based on the amount of water to be used (usually to be metered). To ensure each resident can accesses to the tape-water, the price is usually set lower than the cost, only considered past of the cost of taking and treating water.

Each year the government should subsidize it in order to keep their operation. Although the low rate of price guarantee most of residents access to the tape-water, but standing in other point, this price policy also encourage inefficient water use and meaningless for public environmental-friendly awareness arising.

Also the arrangement place too much financial burden on the government, and made the operation fund for water company is not so abundant, therefore the maintaining of the water transfer network will be lagged, and led more water lost before the water reach consumers tapes.

In recent years, some reform was made on the tape–water price. The price of tape water is raising in some cities, and the sewage discharge, which occupy public sewer system, is also be charged and include into the tape-water price based on the amount of water each household use. But irrational pricing policy still exists. For industry using tape–water, the price rate will be low down if the amount of water exceeds a certain quota.

In addition to the measures mentioned above, water saving techniques is widely used. At present, the emphasis of water saving in Changjiang River Basin is put on the modification of irrigation projects and channel lining as well as the irrigation pattern changing. In cities, several water saving measures are also be taken, such as: water saving facilities used in public toilets; domestic recycling waste water for irrigation; recycling water using within enterprises, etc.

Furthermore, to minimize the water lost from the tape net, some provincial administrative departments give supports to the enterprises to do water balance testing work, and it contribute a lot to those enterprises for water saving.

SUGGESTIONS FOR DEMAND MANAGEMENT

Rising the water price to promote efficient water use and sustainable development of water resources The current water prices, either water resources fee or tape-water price or water charge for water supply projects, are set too low due to several causes, and the rate for same type of charge are set un–flexiblly, also the payment from users can not recover the cost.

All kinds of irrationals in water pricing can not promote the initiative of water supply sector, and it is no help for improving the public awareness.It is suggested that water prices for all types can be risen based on current water supply and managing system.

The price rates charged by water supply project can be set at different levels, such as: subsidized price, planned price and non-planned water price (water supply project contracts the amount of water used monthly with the users, and give a profit price to it. In case that the water used exceeds the contracted one, the higher price rate will be implemented); seasonal price (water price will be changed with the wet and dry seasons), etc.. The water supply managing units will shift to cost-recovering and financial independence step by step.

For tape–water price, it is suggested that full cost-recovery with low benefit pricing rate could be implemented in the economic well-developed regions. For water resource scare regions, the water resource fee should be risen up to some degree, especially for ground water with good quality withdrawing.

Educating people more frequently by multiple ways to improve public awareness. Water saving is always be encouraged in China, and it is considered as one of basic national developing strategies. In each 'Water Week of China' many educating activities are held and presented to the public through televisions, newspapers and broadcasting etc. by water management sectors.

But the education is not only the responsibility of water sectors; it should involve all stakeholders, including enterprises, companies, schools and each householder. Enhance current laws and regulations, and speedy up the step of stipulating of new related laws and regulations Due to the imperfect of some present laws and regulations, there are difficulties existing in implementing those laws and regulations in practice.

Now the Water Law is under amendment. It is suggested that regulations for water fee charged by water supply projects should be amended and enhanced soon according the implementing difficulties existing in practice. Several regulations or laws, which is under consideration, should also be enhanced and published soon.

Reform current water supply system, privatization of water supply can be tried in some region. Now private tape-water supply companies, which were invested and operated by foreign companies, have already put into use in Guangdong Province and Shanghai City. It seems that the water use efficiency of those companies is much better than the ones under the traditional operation.

To ensure water consumers can get good quality service and protect the public health, the government and private companies make contracts on the guarantee degree of water supply and water price, moreover the government has the rights to check the water quality without informing the private companies at anytime. It is recommended that this kind of water supply system could be experienced in more cities and regions.

Favorable policy should give to the utilization of water saving technique The government should stipulate favorable policy to encourage and support the utilization of water saving technique. Grants and soft loans can be supplied to water users that wish to invest in water saving measures, such as: give subsidies to irrigation projects for water saving measures taking; low interest rate loans can be given to enterprises which are willing to invest in water circling use or other water saving measures; water saving facilities should be sale in low benefit or non-benefit.

Demand management has contributed a lot to integrated water resources management in the Changjiang River Basin, but it still has a long way to go. Water resources manager should help and encourage water users to use water saving measures and restructure water use pattern according to the local water resources conditions, while they implement economic and legal instruments for demand management.

OBJECTIVES USING ECP OPTIMIZATION

To design a P&T system, we need to determine well locations and flow rates, which can significantly affect the system performance and cost. Numerical simulation models for flow and transport are often used to evaluate potential P&T system designs.

The simulation model is executed repeatedly to simulate different pumping scenarios. Combining an optimization method with the flow and transport simulator enables optimal pumping scenarios to be selected.

ECP AND P&T OPTIMIZATION

The authors have conducted research on the simulation-optimization approach for over a decade. Based on the evaluation of many different optimization methods for groundwater remediation system designs, we have developed the ECP method. ECP is an extension and improvement over the outer approximation methods that have previously been reported in the literature.

In ECP, new data structures and cutting algorithms are employed to reduce the computational restrictions. ECP has been tested on a variety of published optimization results to find equivalent or improved solutions, and is substantially faster (typically factors of around 100 and greater). ECP does not require a specific flow and transport simulation model be used. It can be wrapped around any flow/transport simulation code. Some of the simulation models that we have used include MODFLOW-96. For a transport simulation problem, the size of the problem refers to the number of nodes (or cells) in the model. For an optimization problem, the size of the problem refers to the number of decision variables.

Due to the storage requirements and computational speed, many optimization methods can only solve medium-sized problems. Therefore, domain-specific insights may be used to select the decision variables (potential well locations) or to reduce the problem size. ECP generally can tackle linear and nonlinear optimization problems of over one hundred decision variables. When the simulated annealing method is used with ECP in tandem, the combined method can tackle optimization problems having thousands of decision variables.

ECP is practice-oriented optimization software. That is, we have investigated a variety of site-specific goals that affect management and design decisions for P&T systems and have built them into the objective and constraint functions. Typically, these goals include the remediation objectives, the anticipated timeframes, initial construction costs (including well, pipeline, and treatment plant construction), treatment costs, and annual operation and maintenance costs. The remediation objectives are often a compromised result of multiple stakeholders.

In ECP, the remediation objectives of a P&T system might include:

- *Containment:* The objective is to prevent the plume from exceeding target concentrations at specified regions within specified time frames. Stakeholders specify Regions–Of–Interest (ROIs). In two dimensions, the ROI may be specified as lines, or areas. In three dimensions, the ROI may be specified as lines, surfaces, or volumes.

- *Points-Of-Compliance:* POCs are finite sets of monitoring locations at which concentration values (and perhaps other performance measures) are used to determine whether a facility is in or out of compliance with a record of decision.
- *Mass removal or cleanup:* The objective is to reduce the contaminant concentrations of the entire site or portions of the site (in this case, ROIs are volumes) to be below specific values (frequently the maximum contaminant level, MCL) at and/or after the specified time horizons.
- *Hybrid:* The objective is mixed containment and cleanup. Typically, the containment objective may be the first priority and the cleanup objective may be the second priority. Containment and cleanup objectives are often described as conflicting (or competitive) objectives, which may lead to conflicting designs. By performing optimization, it is possible to develop an integrated design in which containment and cleanup are complementary.

PROBLEM STATEMENT

In this paper, ECP is applied to a P&T design problem in Massachusetts. In 1994, The site was added to the Superfund National Priorities List (NPL) due to the discovery of groundwater contamination at the site. The primary contaminants of concern are solvents (TCE and PCE) found at depths between 30 and 60 feet below the ground surface. Groundwater flows from the site towards several municipal well fields and to a lake that borders the site on the south. TCE is the contaminant being modeled in this study.

A three-dimensional MODFLOW-SURFACT model of the site was prepared and provided by HydroGeoLogic, Inc. The model has 70,644 cells with 116 rows, 87 columns, and 7 layers. The initial plume concentration was for the year circa 1998 and flow is steady state (except for the remediation-induced flow). The initial maximum concentration at the site was 750 ppb. A dual domain approach is used in the model to represent the retarding mechanisms involving both equilibrium and kinetic adsorption processes. The management time horizon at the site was limited to 27 years, of which the active cleanup time horizon was limited to the first 10 years, followed by 17 years of monitored natural attenuation. Two existing (as-built) wells are pumping with total flow rates of 12902 cubic feet per day (cfd) or 67 gallons per minutes (gpm). Mathematical optimization was applied to find well locations and extraction flow rates that achieve the given remediation objectives while minimizing the total accumulated cost. The target TCE concentrations are 5 ppb.

Two ROIs and a set of POCs were specified:

1. ROI A is described by two line segments at the northwest property boundary. For this 3-D model, ROI A is formed by two planes and has 371 grid points.
2. ROI B is a volume formed by extending the area, Case 3 vertically through all layers. It has 3969 grid points.
3. The set of POCs consists of 9 points of compliance. The TCE concentration at a POC is the maximum value along the vertical direction over all model layers.

We apply the ECP optimization method to the P&T system design problem using four different remediation objectives. Due to the budget and treatment capacity constraints, the maximum number of remediation wells at the site cannot exceed 4. This constraint applies to all design cases.

These 4 optimization cases are listed below:

- *Case* 0: As-built design consisting of 2 existing remediation wells is shown for comparison purposes.
- Case 1: The objective is containment at the northwest property line (ROI A).
- *Case* 3: The objective is maximizing the mass removal over the site (actually over the entire modeled volume) and achieving compliance at the 9 POCs. In this Case, we required that the 2 as-built remediation wells be used.

- *Case* 4: The objective is containment at the northwest property line (ROI A) and mass removal.
- *Case* 5: The objective is containment at ROI A and compliance at POCs. In this Case, we required that the two as-built remediation wells be used.

RESULTS AND DISCUSSION

Optimization may introduce new actions or modify existing designs to reduce risks to human health and the environment, reduce operating costs, and/or shorten cleanup time associated with the remediation. Optimization may also suggest feasible remediation objectives within a given time horizon.

The results of the as-built well configuration and the optimization for the four different remediation objectives are summarized in Table. The plumes at the end of 10 years. The transport simulation shows that the as-built design does not achieve the desired remediation objectives.

In Case 1, two wells are selected out of 50 candidate well locations and pump at a total flow rate of 45.2 gpm. While these two wells successfully achieve the target containment at ROI A, the contaminant mass is not effectively removed. For this case, it may not be possible to close the site at the end of remediation period due to the residual contaminant mass.

Case 2 removes as much contaminant mass as possible from the site while attaining the target concentration values at the POCs. Recall the added constraint that the two existing wells must be used. Table shows that this solution yields the highest contaminant mass removal.

Cases are hybrids involving containment at ROI A together with mass removal over ROI B (Case 3) and with POCs. Both Cases accomplish their objectives. Note, however, that a greater amount of contaminant mass remains in the subsurface, even though it pumps significantly more that the Case 3 solution. This difference is caused by two factors: (1) Case 3 used mass removal over ROI B while Case 4 used compliance at POCs, and Case 4 also required that the two existing wells be used. Table shows that, as a side-effect, Case 3 also satisfies all the POCs.

Table. Description of Optimization Cases for Different Remediation Objectives.

TCE Results (Jan 1998-Jan 2025)	CASE 0– As-Built	CASE 1– Containment at ROI A	CASE 2 – Mass Removal & POCs	CASE 3– Containment at ROI A & Mass Removal at ROI B	CASE 4 – Containment at ROI A & POCs
Number of simulations	1	122	128	108	131
Total Number of Wells	2	2	4	3	4
Pumping, CFD Total Pumped over all years	7325 5577 Tot. = 352.5 MG	4600 4100 Tot.= 237.7MG	9600 6260 4577 9600 Tot. = 820.6 MG	9877 9878 5577 Tot. = 692.1 MG	9600 4577 9600 5577 Tot. = 802.0MG
No. of pts > 5ppb at 10 yrs. at ROI A (out of 371 pts.)	40	0	10	0	0
No. of pts > 5ppb at 10 yrs. at ROI B (out of 3969 pts.)	499	1245	48	0	26

TCE Results (Jan 1998-Jan 2025)	CASE 0– As-Built	CASE 1– Containment at ROI A	CASE 2 – Mass Removal & POCs	CASE 3– Containment at ROI A & Mass Removal at ROI B	CASE 4 – Containment at ROI A & POCs
No. of pts. > 5 ppb at ROI C (out of 10,092 pts.) – 10 yrs – 27 yrs	1026 537	930 469	176 0	240 0	280 62
No. of pts. >5ppb at 10 yrs. at POCs (out of 9 pts.)	4	6	0	0	0
Max. Conc., ppb – 10 yrs – 27 yrs	24.9 14.3	82.9 41.4	15.1 3.4	8.1 4.1	14.0 5.8
(Mobile) Mass Removed, kg In aqueous phase—10 yrs In solid phase —10 yrs	5.4 7.2	4.5 6.0	6.0 8.0	5.9 7.8	5.6 7.3
(Immobile) Mass Removed, kg. In aqueous phase—10 yrs In solid phase—10 yrs	8.1 10.7	6.8 9.0	9.0 11.9	8.9 11.8	8.3 11.0

EVALUATION AND OPERATION POLICY

The chain of mountains in the east of the Mexican Republic called "Sierra Madre Oriental" is the product of a large and very thick sequence of intercalated maritime and continental sediment deposits formed in the Mesozoic era, at the end of the Jurassic and during the Cretassic periods. It is clear that the intercalation took place in alternating sinking and emerging phases of the continent. In the early tertiary age these sediments suffered strong compression stresses which folded and raised them to elevations of thousands of meters over sea level. Later these great rock masses suffered tension stresses with the resulting faults, horsts and graven, the erosion of the soft and elevated sediments with their further deposition at lower levels.

Maritime sediments deposited in shallow waters formed mostly permeable limestone's with dissolution channels (karst) originating in the fossil nodules while deposition in deep waters formed mostly compact and sometimes clayey limestone of low permeability. The alternation of permeable and impermeable strata and its subsequent folding and faulting facilitated the existence of large water deposits in permeable zones contained by impermeable ones in syncline structures.

The northeast part of Mexico is an arid region with very variable extreme temperatures (registered -13°C in winter and 39°C in summer). Mean annual rainfall goes from 200 mm in low continental valleys to 1000 mm in highlands nearer to the Gulf of Mexico. Most populated centres are in the lowlands while highlands are dedicated to weather dependent agriculture with some irrigation by wells and small impoundment. Water demand for human consumption is high in the cities and the resource scarce; this is why nearly 40 years ago the perforation of very deep wells began, searching to exploit the referred permeable limestone.

Results were in many cases favourable with good quality water and high transmissivity of the formation. Nevertheless, piezometric levels were very sensitive to prolonged exploitation and to heavy rainy seasons, which indicated low storage capacity.

In this study it was possible to access registers of monthly observations of piezometric levels and extracted volumes of several wells whose water is used to provide potable water to the city of Saltillo with 700,000 inhabitants. The largest record has 17 years with some missing intervals. The objective was to determine the sustainable rate of water extraction for a well field located in a specific structure (bottom or side of a syncline).

CONCEPTUAL MODEL

The evolution of registered water levels was first examined. It was noted that wells belonging to the same structure have very similar, parallel evolutions even when they are apart as much as 10 km; instead, when the wells are in different synclines, even being very close to each other, their water level evolutions could not be correlated. This suggests that each structure conforms a reservoir which is fairly independent from adjacent ones since they are separated by impermeable layers.

These reservoirs must receive their recharge from rain and surface flow over the surface of permeable layers and have their natural discharge as flow to other formations and also as springs.

The nature of their recharge must be very variable according to the rain regime while their discharge shall change fairly slowly. In natural conditions and in the long run discharge should equal recharge. Pumping from the reservoir introduces a perturbation whose nature will be analysed in what fallows.The hypothesis was made that these reservoirs could be treated as linear ones, that is to say that their natural discharge is proportional to their storage.

MATHEMATICAL MODEL

A quantitative expression for a linear reservoir is given by the continuity equation which for a given time interval is:

$$(V_i - V_{i-1}) / Dt = I_{i-a} (V_i + V_{i-1})/2 - B_i$$

Where:

- V_i–Volume stored at the end of the interval i

- I_i–Volume of water infiltrated from time i-1 to time i (Infitration rate)
- a–Constant that defines the natural discharge as proportional to storage.$[T^{-1}]$
- B_i–Volume of water pumped from time i–1 to time i. (Pumping rate)
- Δt–Time interval

If one considers the subsurface reservoir having a constant plan area and a constant storage coefficient the stored water is expressed as:

$$V = H\,A\,S$$

Where:

- A–Plan area of the reservoir
- S–Storage coefficient of the permeable formation
- H–Piezometric elevation over a base level

On the other hand, infiltration should be related to the amount of rain falling upon the surface of the permeable formation but also to the antecedent humidity of that surface. It has been shown in previous studies that surface run-off and infiltration are a non linear function of precipitation that may be expressed as:

$$I_i = (b\,Pi + c\,P_{i-1}\,P_i\,)\,A_c$$

Where:

- b–Linear infiltration factor for the rain from time i-1 to time i
- P_i–Rain height from time i-1 to time i $[L\,T^{-1}]$
- c–Factor that multiplies previous rain to form the antecedent humidity index. $[L^{-1}\,T]$.
- A_c–Capturing area of the permeable surface of the formation. $[L^2]$

Substituting equations one obtains:

$$H_i\,A\,S\,(1/\Delta\,t{+}a/2) = H_{i-1}\,A\,S\,(1/\,\Delta\,t{-}a/2) + A_c\,b\,P_i + A_c\,c\,P_{i-1}\,P_i - B_i$$

which divided by A S (1/ Δ t +a/2) gives:

$$H_i = d_1\,H_{i-1} + d_2\,P_i + d_3\,P_{i-1}\,P_i - d_4\,B_i$$

Where:

- $d_1 = (1/\Delta t - a/2)/(1/\Delta t + a/2)$ dimensionless;
- $d_2 = A_c b/(A S (1/\Delta t + a/2))$ [T];
- $d_3 = A_c c/(A S (1/\Delta t + a/2))$ $[T^2 L^{-1}]$;
- $d_4 = 1/(A S (1/\Delta t + a/2))$ $[T L^{-2}]$;

To obtain the parameters d_i one depends highly on having trustworthy observations of pumped volumes and of water levels over the largest time interval possible that includes rainy and dry years. For the case studied these variables are measured in general monthly, so that the chosen time interval in the above equations was one month. It should be beard in mind that parameter "a" is linked to the time interval, its value being actually,

$$2(1-d_1)/(1+d_1)\Delta t.$$

DYNAMIC SOLUTION APPROACH

If an error term is added to equation 5, there is a procedure to determine coefficients di in that equation; that is the dynamic least squares fitting by the Kalman filter which also allows to define tendencies of the parameters if they are not constant.

This procedure was applied to several syncline structures of the region under study obtaining very similar results in all of them: d_1 is smaller but very close to unity and fairly constant with variations of $(1-d_1)$ of ± 10%; d_2 is variable but always around zero with insignificant values (10^{-5}); d_3 is very variable around a mean value with fluctuations of 300%; d4 is constant with fluctuations of only ±0.3%.

Two conclusions may be drawn from the described behaviour of parameters d_i: one is that the absence of tendencies (specially the very low variation of d_4) during the observation time with water level variations of around 50 meters, validates the hypothesis of a constant product AS; the second one is that the great variability of the parameter that multiplies the non linear precipitation term d_3, avoids the precise determination of its value with the procedure employed.

The variability can be blamed to the fact that precipitation is not measured directly over the formation and probably also to the influence of the type of rain that occurs during a month, its distribution in time, its intensity, etc.. This has to be further investigated. With these facts in mind an alternative approach was used to obtain the dI parameters.

INTEGRAL SOLUTION APPROACH

Natural Conditions

As was already pointed out, in the absence of pumping and in the long run, inflow to the subsurface reservoir must be in equilibrium with its outflow; the latter is determined by aV = aASH; if a mean initial value of H is taken, the one before exploitation begins, and is named H_{in}, the integral of the inflow volume over a long time span is calculated as $NaASH_{in}$ where N is the number of time intervals considered (in the present case, months).

On the other side, the inflow integrated over the same time span according to the model may be obtained by cAc S $P_{i-1}P_i$; but cA_c from the definition of d_3 is equal to d_3 AS(1/Δ t+a/2); substituting this value and equating integrated outflow and inflow one gets:

$$d_3 = a\,N\,H_{in} / [(1/\Delta t + a/2)\,\Sigma P_{i-1}\,P_i]$$

Exploitation Conditions

When pumping, one can also integrate equation (1) for the whole known time interval; the total storage change is:

$$AS\,(H_{fin} - H_{in}) = cA_{c\,\Delta}t\,\Sigma P_{i-1}P_i - a\,\Delta\,t\,AS\,\Sigma H_i - \Delta\,t\,B_i$$

Where:

H_{fin}–Piezometric level in month N at the end of the observed pumping time.

Again, substituting here the recharge in the long run by the initial natural discharge one obtains:

$$AS = \Sigma\,B_i / [a\,N\,H_{in} - a\,\Sigma H_i - (H_{fin} - H_{in}) / \Delta\,t]$$

From equations and d_3 and AS depend only of observations

and of parameter "a", so that another condition has to be imposed to the solution and that is to have the minimum square error between observed and predicted water elevations. A direct formulation of this condition was obtained but it did not resulted in a reliable value for "a" (probably due to approximation errors).

So, a search method was preferred; the parameter was given different values to obtain d_3 and AS for simulations until the minimum square error was obtained.

OPERATION POLICY

The model itself indicates that no sustainable pumping is possible unless the reservoir natural discharge diminishes and that is precisely accomplished by pumping as the piezometric level is lowered.

Thus, the extraction policy is very clear: one has to over exploit the reservoir until a desirable pumping level H_{des} is reached and then diminish extraction to a rate that equals the rate of discharge that has been avoided to happen naturally; the sustainable pumping rate is:

$$B_{sus} = a\ AS\ (H_{in} - H_{des})$$

APPLICATION TO SOME AQUIFERS

The described evaluation method was applied to five well fields in structures not related to one another, even when some were close together. The best well field is located on the northern slope of a mountain ridge named Zapalinamé where 12 wells were perforated in a length of 10 km. Their water levels show a definite parallel behaviour in spite of their distance and of having experienced a draw down of about 85 meters. Another good well field is that of Jagüey defile with only three wells which receives flow contributions from the two adjacent structures. The other three fields are located at the nose of ridges that deepen into valley sediments. Their evolution although similar is not of the same magnitude, indicating a less uniform structure than the former.

The main characteristics of each well field are given in table. There, the last two columns refer to the relative mean quadratic error in fitting calculated to observed water level evolutions (MQE) and to the head draw dawn necessary to obtain 100 l/s of sustainable pumping. This last value shows the great difference among aquifers.

Table. Main Characteristics of the Analysed Structures

Well field	Observation interval	Extraction Mm³/year	Draw down (m)	H_{in} (m)	$\Sigma P_{i-1} P_i / N$ (mm²/month)
Zapalinamé	1981-1997	21.0	85.0	120	26,300
Jagüey	1988-1996	5.6	.0	180	77,700
Loma Alta N	1981-1994	6.3	55.0	150	18,400
Loma Alta S	1981-1997	6.7	106.0	150	16,900
Puntas	1988-1997	4.0	60.0	100	73,800
ΣH_i (m)	a	AS	d_3	MQE	DH/100 l/s
14,797	0.010	2.20	5.5e-04	0.0070	

The developed model for karst aquifers in folded lime stone allows the evaluation of its recharge, response to pumping and infiltration and is the basis for its operation policy. Based on the concept of a linear reservoir and the continuity equation the model has parameters with physical significance. They can be determined by integrating the continuity equation for natural conditions and for the pumping time intervals.

The latter requires the observation of water levels evolution and of extracted volumes of water. The sustainable exploitation volume equals the amount of natural discharge that is avoided through lowering piezometric levels. This will affect naturally flows to other formations and springs.

The philosophy that may be drawn from this experience is the same that applies to other more common aquifers like those in unconsolidated sediments. In any case they are natural reservoirs that receive an inflow and have naturally an outflow to other water bodies. Pumping for local use will always affect resources downstream.

FUTURE FRESHWATER DEMANDS

THE RECORD

Understandably, the management of freshwater to meet growing demands has been a top priority challenge throughout history, as witnessed by Persian tunnels, Egyptian canals and Roman aquaducts. The best engineers at the time were involved in these impressive civic works, many of which have lasted until this day.

The creation of infrastructure was for a long time associated with peace-time military engineering, but developed into civilian government branches of travaux publiquesthat have performed admirably in many countries, and still do. The fountains of Rome and the gardens in Versailles testify both to the sophistication of earlier epochs and to the skills of the contemporary hydraulic engineers.

They also reveal the shortcomings, which were mainly a lack of knowledge of medical hygiene. While water was delivered in quantities which exceeded present day per capita consumption many times, the water quality was often inferior and sometimes unsafe, due to intruding wastewater. Epidemics of diseases like cholera took the population in London and many other cities by surprise before it was recognised that they are water-borne. Among the many theories for the decline of The Roman Empire one is the possibility that because the water pipes made of lead, Rome's population was suffering from lead poisoning.

The inherited top-down water management still prevails and delivers, in most cities, water of acceptable quality in the quantities demanded to their supply grids, proving that the engineers in charge of water supply to cities of just about any size are still competent, although the glamour of their profession is lost. However, all is not well, as we are reminded of every 22 March, the World Water Day, and frequently through news items such as today's random example (Le Monde, 12 Oct 00: The city of Toulouse has only 24 hours of reserve in case of a serious pollution of the principal river supplying that city).

The sad fact is that more than a billion people, made up of the marginalised population in the urban slums, and overpopulated rural areas, do not have access to a safe municipal grid for drinking water. The squatters in their shacks surrounding most mega-cities do not show up on any plans and are virtually neglected by all authorities, including the water ones. Andrural water supply is in a sad state in many countries.

LIMITS TO GROWTH

So far the supply of freshwater has been sufficient to meet the growing needs. Concern over the fraction of freshwater that can be harnessed in the hydrologic cycle, is a recent topic, and a a very disturbing one. As late as in 1976 none of Kahn's many scenarios in "The next 200 years" mentions water shortage as a possibility, although he is worried about the effect of water pollution on the environment.

Freshwater was for all practical purposes unlimited, or so we thought until a few years ago. The disturbing news that it is not, became widely known through the inventory by Postel *et al.* They demonstrated that the present world population of six billion is already using more than half of the useable fraction of the total annual run-off, which is 40000 km^3. From the gross total they deducted flood run-off and run-off in thinly inhabited regions such as the Arctic and the Amazonas, which are also too remote for transfer, to arrive at the available global freshwater resource.

Their deductions reduced the reserve from comfortable to a mere 0.8 times the present consumption of 6780 km^3/yr. Evidently, since a linear extrapolation of the demand based on today's per capita consumption meets the available freshwater run-off within 25 - 30 years, something has to be done with the demand as well as the supply before the world runs out of naturally recycled freshwater.

Table. Annual Freshwater Run-off

	Volume km^3	Percent
Run-off	40 000	100
Flood losses	20 400	51
Geographically remote	7 100	18
Accessible	12 500	31
In use	6 780	17
Reserve	5 720	14

It didn't take long for this result to sink in and change our priorities. On their list of problems facing us in the next 25 years, a UN panel of futurologists in 1998 ranked freshwater second, after population control.

With the new insight provided by a simple, but long overdue inventory, freshwater shortage is now considered a greater threat than nuclear war, epidemics, food or energy shortage, and climate change.

The gross figures in Table are supplemented by the joint distribution of run-off and population in Table, but in order to map the risk areas, it is necessary to get down to the regional and local levels.

Table. Joint Distribution of Run-off and Population

Continent	Runoff %	Population %
Europe	8.0	13
Asia	35.8	60.5
Africa	10.6	12.5
North and Central America	15.2	8.0
South America	25.6	5.5
Australia and Oceania	4.8	0.5

GOALS AND ANTIGOALS

An interesting alternative to the prevailing philosophy of maximum happiness has been proposed by sociologists and political scientists concerned with the planning of community development, which is a highly unpredictable process towards noble, but misty ideological goals. Since consensus on ideological goals is all but ruled out, the reverse concept of antigoals is introduced: Even if we disagree on where to go from here, it should be feasible to agree on where we don't want to end up. Instead of maximum happiness we should perhaps strive to achieve minimum unhappiness.

The metaphor is the state ship with a captain who doesn't know which port he is to call. Forced to sail without a destination, the captain must avoid immediate disaster by steering away from shoals and rocks and shorelines. Avoiding these antigoals is a strategy with which everybody on board agrees, and the voyage can continue forever if the ship doesn't hit any antigoal. The classical problems of our profession are nothing but antigoals and need only to be redefined as such: Lack of freshwater, loss of freshwater quality, loss of biotopes and biodiversity, no more environmental and human catastrophes like the Aral Sea, no more tragic resettlements like Sardar Sarovar etc.

DECARTES' ERROR

Why is it then that these consensus antigoals are often so hard to obtain support for? An intriguing answer is suggested by Seip and Wenstøp in a recent study of decision-making. They have tested some of the most important water management decisions in Norway recently according to two criteria: The legitimacy of the decision-makers, and the emotional content of the decision. While the first criterion is obvious and in line with all present trends towards more transparency and democracy, the second criterion is somewhat suspect: What have emotions to do in a rational process? Shouldn't they be banned?

The answer appears to be that its emotional content is essential for a so-called rational decision. Without stirring the right emotions, any issue, no matter how important, may go wrong.

Accordingly, a wise decision, securing support and spawning action, has a positive emotional content. Convincing evidence has come from neurologists that have acquired new knowledge about our neural network and our brains. Their findings undermine Decartes' long-lived concept of our pure spirit which is capable of reason as long as it is uncontaminated by the emotions that rule our bodies.

GETTING ON THE AGENDA

It is not sufficient to define societal antigoals that are easy to agree on, because they define advance with a minimum of risk towards the uncertain future. The difficult part is to get the antigoal on the political agenda before it develops into a dramatic crisis, and here we are in the hands of the practitioners of two metiers: journalism and public relations.

The recipe of the journalists is given by Aubenas and Benasayag in "La Fabrication de l'Information". The book's subtitle is Les journalistes et l'ideologie de la communication, and it reveals how the mechanics of this ideology creates the news we are served. "The work of the journalist", the authors state, "does not often any more consist of reporting the realities of the world, but to select among them what should be presented... Everybody knows today that the newspapers reflect less the reality than the representation they have created". And what they select, follows from their code and its hierarchy of criteria. The normal media focus on the nearby (the law of proximity) is only overruled by the spectacular, the dramatic, the extraordinary. An obsession with transparency, easily understood from the point of view of investigative journalism, is not always effective: "If it doesn't tolerate grey zones, the press is condemned to support less and less the realities". These constraints on access to the media are quite severe. If a problem is distant in space or time, or complicated, it is not interesting. The preferred agenda is local, simple and dramatic.

Alas, the journalists do not run the media alone. They must comply with economic imperatives and accept on one hand paid advertisements, and on the other hand ready made stories and news items that do not require further expensive processing to satisfy the rules. The task of the public relations expert is to translate his client's thoughts into the language of the journalist, making them fit for presentation in the media.

A classical challenge is the pedagogical one of presenting the matter in a way that engages the audience. Here the Sesame Street educational TV series has had a lasting effect, successfully competing with commercial TV entertainment and pioneering edu-tainment, the modern, or perhaps post-modern, version of learning through play. There is a close resemblance between the journalistic prerequisite of drama and the educational use of puppets to convey a message. In general, however, the pedagogical tricks are well known and accepted, while those of the media are less well known. Three educational examples may illustrate the points made above.

Convention on the Non-Navigable Uses of International Watercourses

This convention was adopted by the UN in 1997, after a gestation period of more than twenty years. To become law it must be ratified by 35 nations, however, as of today only 12 nations have done so, and it is an open question whether it will ever achieve the status of a binding law. The legitimacy of the decision-makers is unquestionable, but the emotional content of the decision is bewildering.

The Treaty on Banning of Land Mines

By contrast, this treaty, adopted in 1995, was ratified by a record 110 countries after two years. Clearly its uncontroversial emotional content, dramatically as well as gracefully exposed by Lady Diana, unified the world and secure.

The World Water Vision

When an entire profession, such as the water engineers and scientists, makes an effort to impose freshwater on the world agenda in order to forestall a wide-spread water crisis, this is something unheard of previously. With a massive input generated by some 15000 enthusiastic water experts world-wide, the Vision's message that was presented in the Hague in March 2000, is far from satisfying the basic journalistic rules.

It is global instead of local and complex instead of simple, but it does contain extremely dramatic elements. To make the Vision palatable for the media, an enormous PR campaign was staged. Judged by the criteria of the PR profession, which is counting the number of articles, TV showings, interviews etc, the campaign was a great success. However, its real impact is not measurable, and one can only hope that the vast information diffused world-wide by the 1000 journalists and media technicians present, was not wasted.

The Forum itself attracted more than 5000 water experts, but only about 100 politicians attended, almost all ministers of water resources. Thus the meeting was essentially a gathering of the water community. Clearly, much work needs to be done before the next World Water Forum if the freshwater problem is to be taken seriously by everybody and not just by the experts.

HYDROTECHNOLOGY OR HYDROPOLITICS

If we want to avoid a crisis, it is necessary, but not sufficient, to do our hydrotechnical homework. No matter how good our plans are, they don't convince the decision-makers unless they strike the right sentiments in the community. The world is full of noble causes competing for attention and funding, so a skilful marketing is required as well. The sad reality is that the quality of the presentation is sometimes more important than the technical quality of the project.

The easy way out for us hydrotechnicians is to wait for the crossover point in time when the demand exceeds the supply, and the process becomes crisis-driven. But a less cynical and more civilised approach is to work for a pro-active solution, informing, educating and influencing stakeholders as well as decision makers. In short, we must get involved in hydropolitics and hydrodiplomacy. The decision tools we possess, or are developing right now to promote better hydropolitics, belong to the new and exploding field of hydroinformatics.

It employs communication and information technology (CIT) to improve the quality of hydrotechnical solutions, but it also addresses the equally important process of project acceptance and approval, which is what hydropolitics is all about. The classical decision support developed from industrial operations research and often referred to as water management science for a good example), is popular among water authorities and technocrats as it provides a multi-objective optimisation technique based on decision mathematics.

The emerging hydroinformatics, on the other hand, uses CIT to encourage stakeholder participation, which may introduce ideas that challenge the authorities. With hydroinformaticians within their ranks, the water engineers have taken a decisive step to include hydropolitics on their agenda, albeit in an advisory role as facilitators for both top-down and bottom-up processes. CIT makes non-governmental initiatives feasible, whether by professional, commercial or grassroots organisations. It is an equaliser, making water management more democratic, but also more complex and challenging. Naudascher argued for full compensation to the losers when new reservoirs cause involuntary resettlement, and a voice for all stakeholders in the decision process. Using the same tragic evidence as Naudascher, Abbott goes a step further, arguing that the water professionals associated with projects which seriously degrade the lives of the losers, are accomplices of what he considers crimes against humanity. He sees the democratisation of the decision making process as the next big challenge for the hydroinformaticians, a task that may also restore the public image of our profession.

THE WATER MARKET

Local buy and sell water markets have always existed in many places where there is seasonal water shortage. Predictably, such local water markets will increase in number and size as a result either of increased demand or of decreased supply, and they will include long term opportunities. For seashore sites the market price is limited by the cost of locally desalinated sea water, which depends heavily on the local energy price.

This upper bound defines the water market, which Shuval has named the economic watershed: That geographical area surrounding a water short area where the cost of imported water is equal to or less than the cost of locally desalinated sea water and/or recycled treated waste water. Together with the concept virtual water coined by Allan for the water required to produce food and other commodities, the boundary conditions for a freshwater market are set.

Table. Virtual Water. Typical Water Requirement for Production of Foods in California,

Food (kilogram)	Water (liters)	Food (kilogram)	Water (liters)
Wheat	1,273	Beef	16,193
Rice	2,005	Pork	5,760
Maize	978	Poultry	5,730
Potatoes	147	Eggs	3,740
Sugar	2,731	Milk	971
soybean oil	21,692	Butter	22,274

This market is facing many constraints, ranging from the religious claim for water as a gift of God and the philosophical claim of the right to water as part of human rights, to the simpler arguments that dependence on water import weakens the national self-sufficiency and the national security.

Water Trading

Barter may involve more than two countries, often all countries in shared water basins, and sometimes countries in adjacent basins. Before one can start trading, an agreed allocation scheme must exist. Barter can then improve the value of the allocated water by taking advantage of existing hydrotechnical structures such as storage dams, diversion canals and tunnels, or even the construction of new facilities. An intriguing example of this kind of water trading, which he has given the name 'wheeling', is offered by Wolf. He considers a technically simple diversion of water from Syrian and Turkish rivers to Northern Israel, in exchange of increased Syrian and/or Jordanian abstractions from the Yarmuk before it enters Israel from the East. Some of the diverted water coming down the Jordan River may be earmarked for Palestine's West Bank, and allow West Bank groundwater to be diverted to Gaza. As Wolf observes, water can be transported in a water grid much as electric energy is transported in an electric grid. The system can be designed to allow inputs and withdrawals at numerous nodes and can be operated with payment in cash or in kind, *i.e.* barter. The hard part is neither the hardware nor the software of the water conveyance system, but the hydropolitical decision process.

Technology of Transfer

Canals

The Roman aquaducts demonstrate interbasin water transfer as they gracefully span valleys. The energy source to power the transfer was gravity, but water was also lifted up to the aquaducts by pumping. Modern canals frequently cross valleys through tunnels underneath rather than on bridges above the valley floor, making them less visible, but more efficient.

Pipelines

Pipelines are the common carrier of imported water to urban areas. The pipeline can be built through almost any kind of terrain, either on the ground or buried, and the water is better protected against pollution (and pilferage) than in a canal. Submerged pipelines are feasible and sometimes the cheapest solution. Politically a pipeline on the international sea bed may be a feasible, although expensive alternative.

Tankers

Commodity transport by tankers is common, with oil as the prime example. The chief advantage of ships is their flexibility, as they can call on any port for loading and unloading, given a minimum of terminal facilities. While oil is priced so high that it can afford the heavy tanker transport costs, water cannot, except for disaster relief.

Barges

Towed or pushed barges are used for commodity transport in calm waters, mainly, if not exclusively, on rivers and lakes. Essentially a container, the barge incurs costs considerably less than those of a tanker.

Bags

A spinoff from the oil spill combat technology, the plastic bag designed for water transport resembles the barge in that it is towed, but differs in that it is towed only one way. For the return trip it is rolled up on the deck of the tug boat, thus saving fuel and cutting the round trip time.

Icebergs

The idea of towing large icebergs from the Antarctic was investigated in the sixties and first found feasible, but later withdrawn. The second thought was a scenario in which the big iceberg spawned small icebergs during transit, setting the stage for potential "Titanic"- type disasters.

For hauls up to 100 or 200 km, imports with water bags is cheapest and can compete with locally desalinated water. The exact distance depends primarily on the energy costs for the desalination plant and for the tug, respectively.

SECURING SUPPLY

Since freshwater is the only indispensable and irreplaceable life supply commodity, its continuous supply must be secured. A scheduled stop in delivery cannot last many days before it has serious consequences for human health and welfare, and the impact of a sudden breakdown may be disastrous. Water contamination, as we have seen, spreads epidemics fast, so water quality must be continuously monitored and maintained. (Moreover, the taste of water is important, as the exploding market for bottled water testifies. The closest thing to water wars at present, as observed by the media last year, is the cut-throat competition between the companies producing bottled water).

Legal Claims

When increased upstream abstraction or storage is the cause of water shortage, litigation may provide a solution, although protection by law of water supply is generally not secured, in spite of the much talked about human right to water. Unfortunately, the 1997 UN Convention on the non-navigational uses of international watercourses is not ratified into law yet, and may in fact never be. Nevertheless, that convention, as well as the 1966 Helsinki rules, are used as guidelines for negotiated agreements. Although they are not legally binding, the de facto power of *e.g.* the Helsinki rules has proven to be substantial.

Negotiated Agreements

Litigation is in any case a slow process, so temporary import may be necessary even if a verdict may secure the long-term supply. The more fruitful approach has been to negotiate a water sharing agreement directly among the riparians, as has been achieved even in controversial catchments such as the Indus, the Danube and the Rhine. Minimum in-stream flow and maximum allowable concentration of pollutants are typical key items of the eventual agreement, which may also contain the licensing of reservoirs for flood mitigation and hydropower production.

However, all these rules cannot secure the safe supply of water in basins with periods of draught, and do not eliminate the need for occasional import during extreme hydrological events. Given these severe constraints, it is understandable that water import is a tricky business. Ample storage eliminates most of the risk, but there exist ambitious schemes for continuous import with a minimum of storage. Perhaps the most advanced is the water bag system developed by Nordic Water Supply ASA. With three bags one can in principle deliver water continuously: While one bag is being emptied, the next bag is in transit, and the third one is being refilled. (The return voyage is made quickly with the bag on deck). This sounds like a vulnerable, weather-dependent system, but it can readily be made more reliable by adding storage bags. If storage reservoirs exist at the destination, water import may not be more complicated than, say, grain import, with its easily visible storage silos in the harbour area.

Compared with canals and pipelines which can be closed in case a dispute arises, sea transport is less vulnerable because the supplier can be changed, if necessary, to maintain the supply. The analogy to the grain market is obvious and includes the exporter's and importer's option to subsidise a wanted trade, for political or other reasons.

Virtual Water

The water we drink is a mere 1%, and all our other domestic and industrial needs are only 10% or so of the water consumed in producing the food we need. While we are prepared to pay what it costs to deliver domestic and industrial water, the irrigation farmers in dry climates are unable to pay more than a fraction of the actual water cost of their products.

Yet for many good reasons irrigation farming is frequently supported with subsidised water in these so-calledwater-stressed countries with less than 1000-2000 m3/p yr. Figure gives an example of the role of virtual water in a water short area. When the economic value of water becomes higher for other uses than for agriculture, it becomes politically increasingly difficult to maintain traditional agriculture. The response of the farmers has been to improve the efficiency of their irrigation system and to switch to crops with lower water demand, both in terms of quantity and quality. A promising trend is the development of new salt-tolerant plants. These and other trends in agriculture are motivated by the competition from the market for virtual water.

What emerges is a plethora of opportunities for the supply of freshwater, resulting in an ever more complex supply web of real as well as virtual water. The bottleneck is not so much a global scarcity of water as a scarcity of decision power to choose among so many alternatives, all with numerous unwanted side effects. Hydrotechnology is continually lowering the energy requirement for upgrading vast if not unlimited sources of used water and sea water to any desired quality, thus securing our future demands. For the short term interbasin transfer, including trade, is made ever more feasible. Hydroinformatics has come up with a bag of new decision support tools and is developing object oriented approaches at a rapid rate. In particular, these efforts seem to secure stakeholder participation and enhance bottom-up input.Hydroecology continues to educate us on instream water requirements to preserve a desired water and ecotone biological diversity and aestethic quality. Hydropolitics is aided by a growing acceptance of principles for international water law. However, lacking binding laws, unilateral action in shared basins is still a problem.

GRAVITY CURRENTS IN A DENSITY STRATIFIED POROUS AQUIFER

Artificial recharge the artificial increasing of the amount of surface water entering an aquifer is used for many purposes.

7*The most important of these are:*

- The improvement (purification) of water quality,
- Storage of excess water from wet periods for subsequent use in dry ones,
- Maintenance of ground water levels and
- Disposal of treated unwanted water.

Most of published knowledge of the state of the art of the hydrodynamics of artificial recharge assumes that the water of the aquifer has a uniform density or assumes sharp fresh -salt-water ambient water; see for example Mahesha, Thompson *et al.* Little attention has been given to the case of the artificial recharge of an aquifer where the density of the ground water is linearly stratified. This situation may appear in a coastal region.

The basic objective of this paper is the stratified aquifer hydrodynamics of the artificial recharge by a two dimensional source of strength 2Q units of volume per unit time per unit length (line source), embedded in a porous medium, in a linearly density stratified porous medium (aquifer).It is assumed that the porous medium has a constant intrinsic permeability k and is saturated with water of viscosity μ. It is assumed that the artificial recharge scheme examined in this study, is accomplished by injection from a perforated long pipe in a linearly density stratified porous medium of uniform porosity.

The density of the recharging water is assumed equal to ρ_0, which for simplicity is assumed equal to the density of the ambient fluid at the level of the recharging pipe source. In this we present relationships for the growth of the longitudinal length L(t) of the gravity current, intruding the saturated aquifer,as a function of time. After the entry phase, the recharged water will flow more or less horizontally through the aquifer over extended distances, involving long periods of time. The behaviour of submerged intrusions due to artificial recharge is of interest to hydraulic and environmental engineer, because it is related with the underground water quality.

LENGTH OF THE GRAVITY CURRENT

CONTINUITY EQUATION

The groundwater is stagnant and linearly stratified, so that the density, $\rho(z)$, decreases with increasing elevation z.The porous medium is assumed homogeneous. Two dimensional Cartesian co-ordinate axes are chosen with the z-axis directed vertically upwards. we may assume that the line source of water volume flux is a perforated pipe; the recharge is achieved by applying pressure to the water in the perforated pipe. The volume flux per unit length of the pipe is kept constant and equal to 2Q, it starts at the time t=0 and the density of the recharged water is equal to ρ_0, where ρ_0 is the density of the stratified fluid at the level of the pipe location. The flow out of the pipe in the porous medium spreads horizontally at its neutral level and forms an intruding submerged gravity current in the saturated, density stratified porous medium. It is assumed that the flow out of the perforated pipe at x=0, impinges the ambient porous media and the ambient stratified water, rises and descends, and it produces the initial region of the intruding patch.

It is reasonable to distinguish two regions:

1. The impingement region where the flow out of the pipe establishes the initial condition for the intruding water;
2. The main spreading region, which is outside the impingement region.

It may be argued that entrainment in the impingement region is small, and therefore that approximately the volume of the slug at time t is given by the following equation:

$$\text{Volume of slug} = 2Qt$$

If it is assumed that the typical vertical and horizontal extent of the two dimensional intruding fluid are H and 2L respectively, then (1) gives,

$$HL=Qt$$

VERTICAL MOMENTUM EQUATION

By integrating the vertical component of the momentum equation over the spreading patch (slug) and by neglecting small terms we obtain the physically expected result that the total weight of the slug balances the total pressure force, which acts on the slug surface S, *i.e.*

$$\iiint_V P_s(\bar{X})g\bar{k}\,dV = \iint_S P(\bar{X})\bar{n}(\bar{X})\,dS$$

Where $P_5(\bar{X})$ is the density at any point $\bar{X}$ within the slug and $P(\vec{X})$ is the hydrostatic ambient pressure at any point $\vec{X}$ at the interface of the slug; $\vec{n}(\vec{X})$ is the unit vector perpendicular to the surface, g the gravitational constant and $\vec{K}$ is the unit vector in the vertical direction. Since the hydrostatic ambient pressure depends on the ambient density profile (and the depth), it is clear that equation imposes a relationship between the density of the slug and the ambient density. It is assumed that to the first approximation the density within the slug varies linearly with the depth, so that it is easy to integrate equation in a slug of constant depth H and length L(t) to find the following relationship between the ambient and slug densities:

$$P_{au} + P_{al} = P_u + P_1$$

where P_{au} and P_{al} are respectively the densities of the ambient fluid at the upper and lower interfacial layer of slug, and P_u and P_1 are respectively the densities at the upper and lower interfacial layer within the slug. Since the intrusion layer is neutrally buoyant at the height where it spreads horizontally it is, strictly speaking, the "squeezing" pressure ρ_u, ρ_1 exerted on the upper and lower surfaces of the slug which oblige the slug to spread. For linear ambient stratification and linear density profile within the slug, assuming that the density within the slug of thickness H is given by,

$$P_u + \frac{P_1 - P_u}{H} z$$

where the vertical distance z measures depth from the upper point of the slug, we find that the pressure inside the slug is equal to,

$$P_u + P_u g z \frac{(P_1 - P_u) g}{2H} z^2$$

The pressure inside the slug is clearly greater than the pressure outside, and therefore the xcess horizontal pressure (usually called "buoyancy ") force Fp per unit width which drives the spreading (intrusion) is given by,

$$F_p = \rho' g H^2$$

where,

$$\rho' = (\rho_u - \rho_{au})/6$$

Using the continuity equation (2), equation 7 becomes,

$$F_p = \rho' g Q^2 R^{-3} t^2$$

HORIZONTAL MOMENTUM EQUATION-SCALING ANALYSIS

The methodology that we will follow to find the asymptotic growth rate of the length L(t) with time is based on the balance of the forces, which drive and retard the flow. Similar methodology has been used previously by Hoult, Chen and List Didden and Maxworthy, Lemkert and Imberger. The force, which drives the flow, is only one: the pressure (or buoyancy) force Fp. The force, which retards (or resist) the flow is only one, the drag Fdrag which is exerted by the ambient porous media and the ambient fluid on the intruding fluid (clearly the inertia of the intruding gravity current is negligible). Subsequently we find the scaling of the above-mentioned forces, where the continuity equation (2) has been considered and where L(t)/t gives the typical horizontal velocity U within the intrusion, where t is the time.

We find for the pressure force,

$$F_p = \text{pressure (or buoyancy) force} = O(r' g H^2)$$
$$= O(r' g Q^2 L^{-2} t^2)$$

We assume that the drag force, which is applied to the intruding slug, is due to Stokes drag forces of the slug fluid due the flow around the numerous grains of the porous media. The Stokes force Fs due to the laminar flow around a sphere for small Reynolds number (Re<1), is given by,

$$F_s = 3\pi d \mu U$$

Where μ is the dynamic viscosity, U the fluid velocity and d is the grain diameter. We obtain therefore,

$$F_{drag} = \text{drag force} = O(\mu\ d\ U\ n) = O(\mu d^{-2}\ LT^{-1}\ HL\ ) = O(\mu\ d^{-2}\ L\ Q\ )$$

where n is equal to the number of grains within the volume of the slug,

$$\text{i.e. } n \approx HL(t)/d^3.$$

We consider below the growth L(t) under the balance of the corresponding driving and resisting force. Clearly we have a balance of the buoyancy driving force F_p and the resisting drag force F_{drag}, and we obtain:

$$F_p = \text{pressure (or buoyancy) force} = F_{drag} = \text{drag force, or}$$

$$O(\rho' g Q^2\ L^{-2}\ t^2\ ) = O(\mu\ d^{-2}\ L\ Q\ ), \text{ which gives,}$$

$$L(t) = c_o \left(\rho' g Q d^2 / \mu\right)^{1/3} t^{2/3}$$

where C_o is an experimental parameter. The above analysis predicts the typical horizontal length L(t) of the gravity current increases with time as $t^{2/3}$.Subsequently we describe experiments conducted to test the above asymptotic law.

EXPERIMENTAL PROCEDURE

A Hele-Shaw cell was constructed of two sheets of 1 cm optically flat plate glass approximately 66 cm long and 48 cm height, clamped 0.1 cm apart.The entire cell was embedded in a tank 20 cm wide and 100 cm long. A vertical wall, made of glass, 19.9 cm wide and 48 cm height, as shown, separated the tank in two smaller tanks, tank 1 and tank 2. The left end of the cell was freely exposed in tank 1, and the right end of the cell was sealed with plastic tape except of a small hole at mid elevation. The hole is temporarily blocked, and it is unblocked at the start of the experiment. Tap water and commercial salt was used to stratify the tank.

The tank and the cell was linearly stratified by running layers of successively increasing density in the bottom and then allowing the tank to stand for about 12 hours to smooth out the discontinuities. The actual stratification in the tank was measured with a calibrated conductivity probe. The density in the tank was uniform and equal to the density at the middle point of tank 1.A red dye was also diluted in the tank. The free surface elevation of tank 2 was 1 to 2 cm higher than the free surface elevation of tank. This difference of the free elevation of the two tanks drove the intrusion into the Hele- Shaw cell A measure of the gradient of the ambient density profile at the level of spreading is given by the Brunt-Vaissala frequency N, which is calculated by the relation

$$N = \left[\frac{gd\rho(z)}{\rho(z)}\right]^{1/2}$$

When a uniform density gradient had been established in tank, the hole which separates the linearly stratified Hele–Shaw cell from the tank 2, wȧs unblocked and the dyed water from tank 2 started to intrude between the plates of the Hele–Shaw cell. In some experiments the water in the tank was not dyed and the intruding flow was visualized by dropping dye particles between the plates of Hele-Shaw cell.

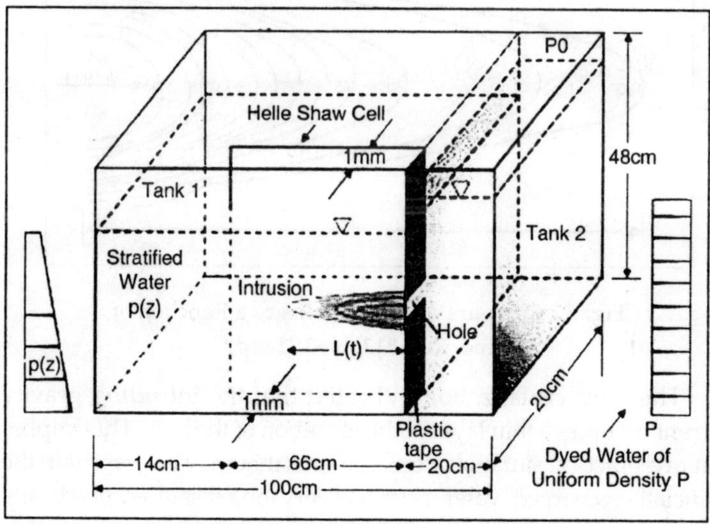

Fig. 7.1 Experimental Configuration (not to Scale) of the Simulation of the Intrusion in a Density Stratified Porous Medium.

EXPERIMENTAL RESULTS

For each experimental run, the length L(t) of the intrusion patch was measured as a function of time. A typical contour of the intruding gravity current as a function of time is shown in Figure 7.1. For each experiment the length L(t) of the spreading slug was plotted in logarithmic scale as a function of time. The raw data of the length L(t) as a function of time t from many experiments are plotted.Best-fit lines are also drawn and the corresponding equations of the fits are printed in the plot.

The best fit equation is of the form L tm,where m is the exponent of time. It can be seen that experimental results show that the length L(t) grows like tm, where m=2/3, in agreement with our theoretical prediction. In addition, the area within each contour was measured.It was found that the area increased linearly with time, so that we verified that the input volume flux Q was kept constant during the experimental run.

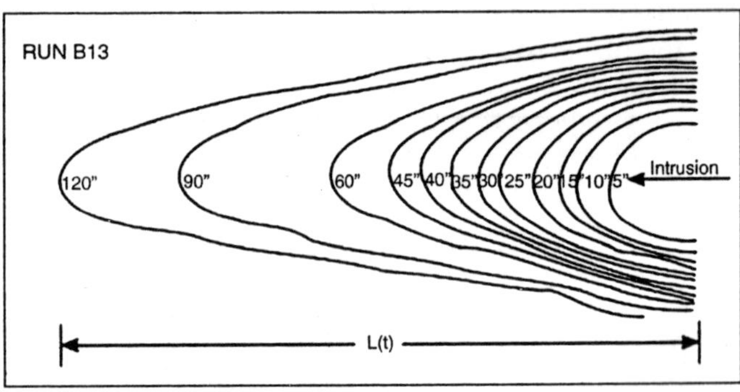

Fig. 7.2 Contours of the Intrusion as a Function of Time; Run B13; N=0.71 sec^{-1}

The flow visualization indicated that the intruding gravity current occupies a thin layer at the elevation of the hole. This implies that in a practical situation, contaminants that may be present in the artificially recharged water can travel very long distances, practically without any further dilution. The experimental results indicate that the length L(t) of the intrusion increases with time as $t^{2/3}$, in agreement with the theoretical prediction.

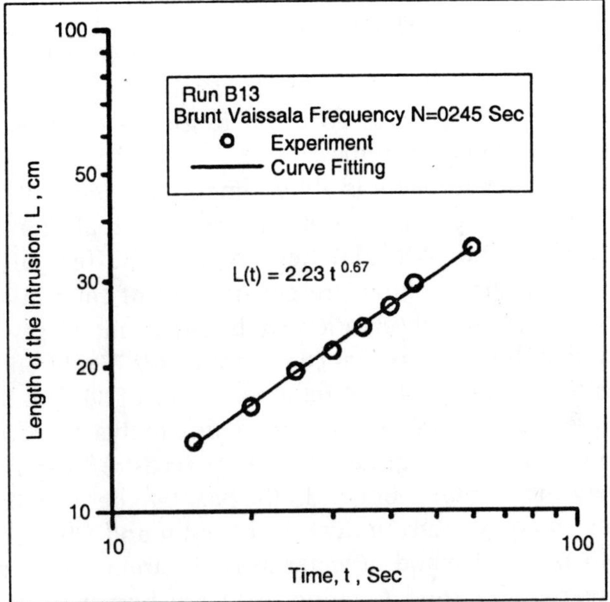

Fig. 7.3 Length of the Intruding Gravity Current as a Function of Time

WATER CONJUNCTIVE USE PLAN

INTRODUCTION

Taiwan is located in subtropical zone, however, due to its peculiar topographic configuration, rainfall is unevenly distributed both in space and time. In the dry season, rainfall is generally scarce in most areas. Under this situation, water resources application is a difficult issue on the island. As a result of continuous economic growth, rapid increase of population, booming industrial development, and highly improved living standard, we see a continuous increase on water demand during the past decades. This increased demand on water renders water supply management all the more difficult than days past. This is especially true in groundwater management.

The major water supplies in Taiwan are surface water and groundwater. Statistics shows that the annual surface water supply is about 11 billion m^3, amounting to about 65% of the total 17.0 billion m^3 water used. The annual groundwater supply is about 6.0 billion m^3, and constitutes about 35% of the total water use. The Chuohsui Alluvial Region straddles the Changhwa and Yunlin counties.

Groundwater resource is a very important contributor to the total water supply, and groundwater is used all the year round in Chuoshui Alluvial Region. The aquifers of the Chuoshui Alluvial Region are constituted of four layers. Information from the observation wells and monitoring wells indicated that the groundwater consumption is 877 million m^3/ year, which is about 20.4% of total water use of 3088 million m^3. The groundwater table draw- down due to this practice of groundwater retrieval is quite serious in the coastal areas of the Changhwa and Yunlin counties. In the past two decades there is a boom in aquaculture industry in Yunlin and Changhwa coastal areas, the demand of water increases tremendously in recent years. The aqua farming industry began to draw groundwater on a large scale, and caused serious land subsidence. It thus stands to reason that, in order to prevent land subsidence and ensure the sustainability of water resources, proper utilization and management of groundwater is an urgent task for the water resources sector.

This paper discusses the characteristics of groundwater and the current key issues of groundwater management in Chuoshui Alluvial Region. Action plans for improving groundwater management and the sustainable use of groundwater are also discussed. It is expected that these management strategies introduced will be effective measures that would curtail the practice of excessive groundwater withdrawal, and gradually, but eventually reduce the amount of land subsidence, and also achieve the goal of sustainable use of water resources.

CURRENT ISSUES ON GROUNDWATER MANAGEMENT

Utilization of groundwater has its advantages and drawbacks. Using groundwater means low cost and readily available water for agricultural, domestic and industrial uses. On the other hand, excessive use of groundwater has imposed some serious problems. During the past forty years, the use of groundwater has created some critical issues on the management of groundwater resources.

The key issue related to groundwater management in Taiwan is that there is no bona fide management body for groundwater use. Traditionally, surface water is managed by the farmer organizations. In this connection, groundwater is only a supplemental water resource for the farmer organizations that distribute water to the farms. Under this system, groundwater management is wholly neglected during the last several decades. By legislature, the direct management authority for groundwater is the local government of each county, and there are commonly just a few clerks who perform the chore of registration only.

Under the Water Law, there is no need to enter a registration in the local county government if the use of groundwater is less than 144 m^3/day. This leads to loss of information, albeit the groundwater drawn is of small quantity in individual case, but the total cumulative quantity becomes quite substantial but is unknown. Another factor is that the farmers and private industries do not really understand the importance of groundwater cycle and the regulations of groundwater management. The combination of these factors explains why groundwater management is at best lax and ineffective, and sustainability remains a distant goal yet to be achieved. After forty years, the results show that only about 10% of water wells have been registered in the local county government.

The second issue related to groundwater management in Taiwan is that in the past forty years there were no effective groundwater monitoring systems. Recently, this situation has been improved by installation of monitoring systems in the Chuoshui Alluvial Region, Pingtung Plain Region, Chaiyi–Tainan Plain Region and the Lanyang Plain Region. The rest of the groundwater basins still do not have modern monitoring systems. Except Taipei basin, most groundwater basins do not have water meters on their wells. Hence, the actual groundwater consumption can only be estimated indirectly.

The third issue related to groundwater management in Taiwan is the incessant draw down of groundwater table and subsequent land subsidence. During the last forty years, groundwater is over drawn in most coastal plain aquifers because of the increased water demand and the insufficient surface water supply. The consequences of excessive groundwater use are the continuous draw down of groundwater table and the consequential land subsidence due to the consolidation process. Figure 7.4 shows the extent of land subsidence of the Chuoshui basins respectively. The consequences of land subsidence are drainage problems on low land, beach erosion and high risk of dikes on the coastal areas.

WATER CONJUNCTIVE USE IN CHUOSHUI ALLUVIAL REGION

In the Chuoshui Alluvial Region, the Yunlin Offshore Industrial Estate is developed from 1994 to 2020. The first phase is to develop 2200 hectares of land for petrochemical industry use. The Mailiao Project of the Formosa Plastic Group is scheduled to be completed from 1994 to 2000. The full extent water demand for industrial use is 300000 m^3/day. The second phase development of the Hsing Hsing Zone started from 1998.

A total of 1000 hectares of land is developed for basic industry. The estimated increased amount of industrial water demand is 220000 m^3/day in 2009. The third phase development will be finished in 2020. The increased amount of water demand will be another 340000 m3/day. Besides the Yunlin Offshore Industrial Estate, there are other industrial developments in this region; these include the expansion of Touliu Industrial Park, the Yunlin High-Tec Park and the Changhwa Shuiwei Industrial Park, etc.

It is foreseen that population growth, economic boom, and raised living standard will be future phenomena in this region consequent to the industrial development. In keeping with this promising industrial development, the total water demand will be increased up to 700000 m^3/day in this area in the year 2010. The increased water demand curve is shown in Figure 7.4.

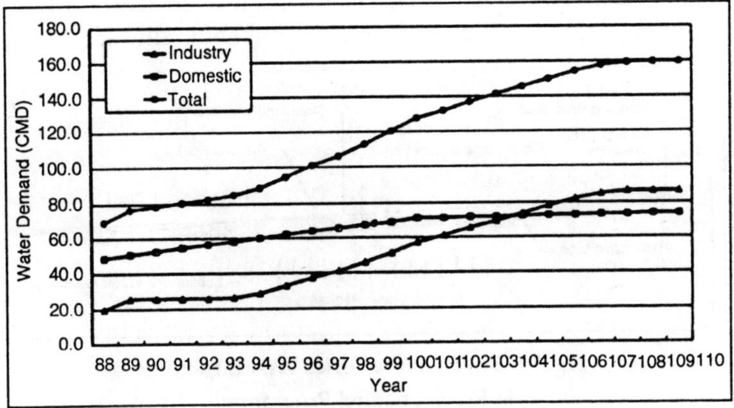

Fig. 7.4 The Water Demand Curve for Chuoshui Alluvial Region

Three stages of water supply schemes are implemented to cope with the rapid development in order to satisfy the increased water demand. The core concept of the conjunctive use is to satisfy the water demand by conjunction of the safety yield groundwater and the available surface water in a basin. In other words, the plan of conjunctive use of surface water and groundwater is to control the groundwater-pumping rate, and supplement by the surface water supply scheme for the future sustainable use of water resources.

The conjunctive use of surface water and groundwater plan is currently being conducted in the Chuoshui Alluvial Region. Figure 7.5 shows the water supply schemes of the Chuoshui Alluvial Region on the conjunctive use bases. Based on the geological formation of the Chuoshui.

Alluvial Region, the artificial recharge programme is to evaluate and select suitable places in the alluvial fan deposits areas for groundwater recharge. During the wet season from May to October, surface water is the main water supply, whereas during the dry season, from November to April, the water supply mainly comes from reservoir and groundwater wells. In the wet season, the residual water is diverted to the recharging pond, or recharging lake, for groundwater recharge. Figure 7.5 illustrates the idea of the series weir structures in river channel programme at work in the Chuoshui Alluvial Region.

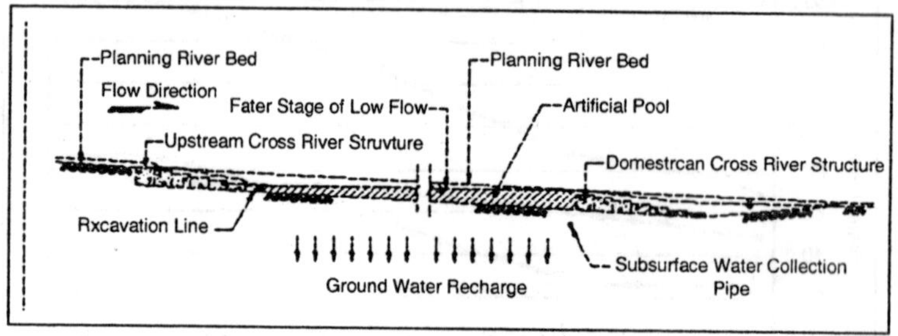

Fig. 7.5 Sketch of the Series Weir Structures in River Channel Programme

The first stage is construction of the Chi–Chi Project. Major multipurpose functions of the Chi–Chi project include steady supply of agricultural water, meet the increased demand of 200000 m^3/day for the domestic water supply, and fully supply the 860000 m^3/day of water for industrial use in the wet season. During the dry season, the supply of the 300000 m^3/day industrial water demand is reallocated from agricultural water supply. The second stage scheme is to implement the Yunlin Reservoir Project, the Hushan and Hunan Reservoir. This reservoir provides 54 million cubic meters storage capacity for surface water during wet season. The project will be completed in 2009. The water supply will be 694000 m^3/day in conjunctive use with the Chi–Chi Project. The industrial and agricultural water supply will be well stabilized upon completion of this reservoir. During this stage, the Mailiao Lake Project will also be implemented. The capacity of water supply from Mailiao Lake Project will be 30 million m^3/year. The Channeling work of Chuoshuichi and artificial recharge project will also be implemented during this phase. An approximate 30 million m^3/year reduction of groundwater consumption will be realised through this conjunctive use scheme.

Modern groundwater monitoring systems are now installed in the Chuoshui Alluvial Region. 188 monitoring wells were installed in its 4–aquifer layers. Under this monitoring system, groundwater tables are measured on the daily bases for the Chuoshui Alluvial Region from 1994. The data are entered onto the data log every day and analysed every month. The information of groundwater tables is published every year, and is available for use by related organizations and interested parties. The information of groundwater table is also transferred to the local county for the purpose of groundwater management.

The third stage of this water supply scheme is to improve groundwater and surface water conjunctive use, to increase groundwater management and to reduce groundwater consumption. The programme of encouraging reasonable water consumption will also be installed in this area for the purpose of reducing extra water demand. Under this stage, the Juifeng Reservoir Project will be implemented in accordance with the economic and social development plan for the Chuoshui Alluvial Region.

CONCLUSION AND SUGGESTION

Groundwater is an important resource in Taiwan. On this island, almost 35% of water supply are from groundwater. The utilization of groundwater has its advantages, such as providing cheap water for domestic, agricultural and industrial uses. However, serious land subsidence problem occurred in many coastal zones due to inadequate withdrawal and excessive extraction of groundwater in these areas. During the last forty years, groundwater use in Taiwan is not sustainable in several areas; this culminated in high cost to the society and future generations.

To improve groundwater management in Taiwan and mitigate the consequence of non-sustainable use of groundwater, several measures are installed and planned on two pilot groundwater basins, the Chuoshui Alluvial Region and Pingtung Plain Region. The information of groundwater characteristics acquired through these measures will be transferred to the local county government, provincial government and central governmental agencies for the purpose of groundwater management.

The conjunctive use of surface water and groundwater is the best solution for increasing groundwater recharge and sustainable use of water resources. There are three major strategies of the conjunctive use, the artificial recharge programme, the series weir structures in river channel programme, and reservoir construction. These strategies are now operating in full throttle in the Chuoshui Alluvial Region and the Pingtung Plain Region. Similar programme will be developed on other groundwater basins eventually. Following establishment of conjunctive use of surface water and groundwater policy, groundwater use in Taiwan will be sustainable in the future.

Land subsidence in the coastal zones is a serious problem in Taiwan. After the urbanization development together with water supply scheme, the groundwater use pattern will be improved in the future on the coastal areas. If the development proved to be a correct approach, land subsidence in the coastal areas will be eventually curtailed, and the water resources will progressively be sustainable in the future.

GROUNDWATER UTILIZATION AND AQUIFER

INTRODUCTION

The rapid population growth and the development of aquaculture and industry during the past three decades have increased water demand from 10 billion m^3 in 1961 to 15 billion m^3 in 1971, and to 18 billion m^3 in 1992. A water demand of 21 billion m^3 is expected for the year of 2010, if the projected agricultural, industrial, and domestic requirements are correct. In addition, due to the demand for water supplies has risen, the advantages of lower costs, easier accessibility, and more stable quality of groundwater have encouraged people to use it. The annual groundwater withdrawal in Taiwan was estimated to be 3.2 billion m^3 in 1975, 4.1 billion m^3 in 1983, 6.3 billion m^3 in 1988, and 7.1 billion m^3 in 1991. The annual groundwater recharge, however, was estimated to be only 4 billion m^3. Over-pumping of groundwater by various water users, such as the extensively developed aquacultural industry in the coastal area of southern Taiwan, has lowered the water table quickly in this area. Consequently, the declining water level has resulted in the consolidation of alluvial sediments and caused land subsidence.

However, the lack of reliable hydrogeologic data makes it difficult to quantify and regulate the groundwater use and control land subsidence. In 1992, the Department of Water Resources (DWR) was requested to initiate a plan entitled "Groundwater Monitoring Network Plan in Taiwan,". Since then, the DWR has taken the responsibility for planning, supervising, and raising funds for this plan. The main objective of this plan is to collect data at regional scale that would be ultimately used for the planning and management of groundwater resources in Taiwan. In 1993, in certain areas, land subsidence only became worse. In Nov. 1995, the "Land Subsidence Prevention and Reclamation Plan (LSPRP) was approved by Executive Yuan.

GROUNDWATER MONITORING NETWORK PLAN IN

The Groundwater Monitoring Network Plan in Taiwan covers nine groundwater basins as well as The Penghu Island and Hengchun plain areas. The calls for 517 hydrogeologic survey stations and 990 groundwater monitoring wells. Because of financial constraints, this plan is to be accomplished in three stages over a period of 17 years. The total estimated implementation cost was NT$ 6.4 billion(around US$ 200 million).

The current plan aims at developing a reliable groundwater monitoring network in Taiwan to collect long-term groundwater data. The scope of the current plan consists of six items. The Groundwater Monitoring Network Plan in Taiwan has been carrying on since 1992. By the end of July, 1998, 145 hydrogeological survey stations and 315 monitoring wells were constructed. The plan is focused on the Choshui River alluvial fan and Pingtung plain during the first stage.

The preliminary hydrogeologic systems, the occurence of aquifers and the locations of recharge areas have been identified. The groundwater hydrology, aquifer characterization, and groundwater flow pattern were analysed. The results are used for the comprehensive planning and management of the water resources in these two basins.

RESULTS OF THE CHOSHUI RIVER ALLUVIAL FAN

The Choshui River alluvial fan, the largest in Taiwan, is located at the west coast of central Taiwan. It is triangular in shape with the apex at Linnei and divided into north and south parts by Choshui river. The area of the fan is 1,800 km ^{2}. In the fan area, the overall water consumption is 3.1 billion m^3, of which 0.9 billion m^3 is from groundwater. The distribution of groundwater monitoring wells in the Choshui River alluvial fan area are shown in Figure 7.5. Geological drilling and stratigraphic analysis were conducted to interpret the subsurface geology in Choshui River alluvial fan.

The groundwater basin is composed of three aquifers and two aquitards, as shown in Figure 7.5. Aquifer II is the most important water bearing formation of the fan area. Its large grain size suggests that it has high hydraulic conductivity. The groundwater stored in the fan area is generally confined in nature throughout most of the area, except at the upper part near the apex of the fan.

The aquitards consist of clay and silt in variable thickness. From pumping test analysis, the transmissivities and the hydraulic conductivities of the aquifers range from 0.01 to 4.19 m^2/min and 10^{-3} to 10^{-5} m/sec, respectively. The specific yields range from 0.18 to 0.29 for unconfined aquifer. The storage coefficients range from 10^{-5} to 10^{-3} for confined aquifer. The apex of the fan can be conserved for groundwater storage and the groundwater should be extracted out during the dry season only from the middle and fringes of the fan. From long-term monitoring, the groundwater level had gradually declined since 1960, but yet rebounded after 1990. Aquifer II has been over withdrawal all the time. The average amount of groundwater recharge is 1.024 billion m^3/yr. According to the study of numerical modeling, the safe yield, which is equivalent to the natural recharge is estimated to be 0.818 billion m 3/yr. The routine analysis of groundwater quality has reached the conclusion that there is no seawater intrusion in the Choshui alluvial fan.

RESULTS OF THE PINGTUNG PLAIN

The Pintung plain is located at the most southwestern part of Taiwan. It is rectangular in shape with 55 km long and 22 km wide, and encompasses an area over 1,200 km^2. The Kaoping creek is the largest river in Taiwan with an annual run-off of 8.5 billion m^3. The overall water consumption is 1.47 billion m^3 per year, of which 1.04 billion m^3/yr are from groundwater. The locations of groundwater monitoring wells in the Pingtung plain are shown in Figure 7.5. A same investigation and study programme was also conducted in Pingtung plain. The groundwater basin is represented by three aquifers and two aquitards, as shown in Figure 7.5.

Aquifer III is the most important water-bearing formation. Aquifers I, II, and III have cropped out in the Manila trench a few kilometers off the coastal line, as shown in Figure 7.5. The aquitards are composed of clay, leading to a low hydraulic conductivity of the strata. From pumping test analysis, the hydraulic conductivities of aquifers range from 10^{-3} to 10^{-5} m/s, and decrease from the apex to the fringes. Transmissivities range from 0.00003 to 15.1 m^2/min. The specific yields range from 0.01 to 3.2 for unconfined aquifers. The storage coefficients range from 10^{-3} to 10^{-5} for the confined aquifers. The discharge per unit drawdown ranges from 0.004 to 654.5 m^3/m/h. The groundwater recharge areas can be delineated which include the proximal parts of Laonungchi fan, Ailiaochi fan, Linpienchi fan and Lilychi fan.

Groundwater level had been gradually declined since 1979, but yet rebounded after 1989. Aquifers I and II have been over pumped for a long time. The water table fluctuates by as much as 21 m during a given month and 5 m during a given day. The average amount of recharge is 1.11 billion m^3/yr, and the safe yield is 1.0 billion m^3/yr. The seawater intrusion in the aquifers has been studies in terms of hydrogeology, groundwater level, groundwater salinity, and resistivity log. The result shows that the ease of seawater intrusion has benefited from the long extension of aquifer towards the Manila trench.

LAND SUBSIDENCE PREVENTION AND RECLAMATION PLAN

The LSPRP has two major goals:

1. To prevent over-pumping of groundwater, thereby alleviating land subsidence, and
2. To make reasonable use of land and water resources, thereby reducing physical damage and social costs. To evaluate the success of the Plan, five quantitative objectives are proposed.

They are:

- The total aquacultural area will be reduced from 52,000 hectares to 22,000 hectares.
- The number of illegally constructed wells are to be reduced by 50 per cent in the areas, where during the past three years, the cumulative land subsidence has exceeded one meter and in the areas where the average rate of subsidence has exceeded 15 centimeters per year.
- Groundwater extraction will be reduced to the safe yield level.
- There will be no further increase in land areas, where the groundwater is below the mean sea level, and to reduce such land areas by 50 per cent.
- Land subsidence in 50 per cent of the affected areas will cease.

Evaluation of the Lsprp

Based upon actual accomplishments of the five quantitative objectives, it is obvious that the LSPRP has not been fully successful. In fact, many obstacles have been encountered during the execution, which have slowed down the progress and hindered the achievements.

Among the obstacles, the following four have been identified as the most serious:

- Transforming of the subsidence area from aquacultural use to industrial use has not been successful due to the high cost and the fact that the natural and social-economic conditions are not suitable for industrial use.
- The county-level governments have been facing difficulties in enforcing the groundwater control regulations due to lack of manpower.
- A small number of new illegal wells has been detected during the execution of the LSPRP. Furthermore, it has been realised that, because of the persistence effect, land subsidence will continue regardless.

The "Groundwater Monitoring Network Plan in Taiwan," which was initiated in 1992 is geared to define the goals and implementation procedures of the effective management of groundwater for Taiwan. The preliminary results of the studies have shown that the conjunctive use of surface-water and groundwater is the best managing strategy for the Choshui River alluvial fan and Pingtung plain. The apex areas should be conserved for groundwater recharge, and the other areas of the fan will provide groundwater during dry seasons. The comprehensive "Land Subsidence Prevention and Reclamation Plan" is Taiwan's first attempt to remediate the extremely serious land subsidence problem with all related agencies participating. After the execution of the Plan (by mid 2000) the specified five quantitative objectives have been achieved to some degree. Although some obstacles have been encountered so that the expected goals are not achieved; however, without question, the implementation of the LSPRP has prevented the land subsidence problem from worsening. By assessing the results obtained to date and the obstacles encountered, the Water Resources Bureau has proposed a second-stage plan for another five years. It was expected that there would be difficulties in reaching the quantitative objectives at the end of the second-stage plan. Nevertheless, it is also expected that, after the second-stage plan is fully implemented, land subsidence in Taiwan will be under proper control.

Table. Planned Hydrogeologic Survey Stations and Groundwater Monitoring Wells to be Installed at Different Stages

Stage	Groundwater Basin		Hydrogeological Survey Station	Groundwater Monitoring Well
	Name	Area (km²)	Number of Stations	Number of Wells
Stage I	Choshui River alluvial fan	1,800	77	175
1992	Pingtung plain	1,130	60	148
I	North part of Chianan plain	300	4	9
1998	Subtotal	3,230	141	332
Stage II	Southem part of Chianan plain	2,220	100	212
1999	Hsinchu-Miaoli coastal area	900	48	85
I	Lanyang plain	400	22	40
2003	Penghu Island	106	13	25
	Subtotal	3,626	183	362
Stage III	Taipei basin	380	27	46
2004	Taoyuan-Chungli terrace	1,090	42	85
I	Taichung area	1,180	47	57
2008	Hengchun plain	110	9	20
	Hualien-Taitung valley	930	68	88???
	Subtotal	3,690	193	296???
Total		**10,546**	**517**	**990???**

Table. Major Work Items and Corresponding Achievements of LSPRP by Mid 2000

Major work item	Achievements
Comprehensive land use planning for areas affected by land subsidence	1. Total fishponds area reduced from 52,000 hectares in 1994 to 41,069 hectares. 2. 1,000 hectares of fishponds transformed to artificial recharge ponds and 3. 1,870 pumping wells stopped in chia-Yi county.
Assistance to industry and other productive activities in subsidence area	1. 1,197 hectares of fishponds transformed to industrial use in Yun-Lin county by the enforcement of the Agricultural Land Release Policy. 2. Assistance in using water-saving techniques to 53 factories, resulting a 6 to 20 percent of water conservation
Enforcement of groundwater control and better water resources planning	1. 2,585 illegal wells capped since 1995. 2. Groundwater extraction reduced from 6.28 billion tons in 1995 to 5.73 3. billion tons in 1999. 4. The continuing subsidence area reduced from 915 square kilometers(km²) in 1995 to 697 km² in 1999. 5. 177 hydrogeological stations, 376 groundwater-monitoring wells and 30 land-subsidence-monitoring wells completed since 1995.
Education and promotion	1. Several discussion sessions for local leaders conducted. 2. Several videotapes for the promotion of the LSPRP produced and distributed. 3. A series of workshops conducted on "Sustainable Use of the Land and Long Last of the Home." 4. Advertisements in boards and buses made. TV programs produced. 5. Promotion notes on cards, mats. booklets and notebooks printed

8

Exploitation of Water Supply

INTRODUCTION

The Vrana Lake is situated on the island of Cres, belonging to the group of Croatian islands of the Adriatic Sea - a marginal sea of the Mediterranean (Southern Europe). With its average area of approximately 5,75 sq.km. and volume of approximately 220 million cu.m. of water of exceptionally high quality on otherwise dry area of the island of Cres (the total area of the island is only 405,78 sq. km.), it represents an invaluable resource in both the environmental and economic sence.

The lake is a kryptodepression, with the water level between 16,70 m a.s.l. and 9,11 m a.s.l. on the average 13,13 m a.s.l. in the period from 1929 to 1995. The lake bottom is at 61.3 m below the mean sea level. The average annual rainfall in the area of the Vrana lake is 1064 mm, the mean annual air temperature is +14.8°C, and the annual evaporation from free water surface is 1161 mm. The falling trend of the water level in the lake in the eighties, never recorded before, caused a great concern among proffesuinals and the public regarding the future of the lake, and initiated the beginning of complex research work. In the recent years this trend has been stopped; however the need for protection of the lake is still present in the sense of defining the regime of water pumping from the lake. Because of its position and size, the issue of the origin of the lake water has been attracting the attention of scientists for a lond time. Already in the past century contrary assumptions appeared regarding its recharging - from the mainland or the local island catchment area. The results of all hydrological investigations carried out so far, both earlier and recent, althought not identical in figures and conclusions, speak in favour of the latter assumption. Mechanism of lake's aquifer functioning in way of hydraulic relation between salt and fresh water was analysed in Ozanic and Rubinic. In Bonacci hypotetical hydraulic model with assumption Lake's recharging from the mainland was analysed. The results of these hydraulical investigations confirm hydrological conclusions, so the problem of functioning and protection of Vrana Lake must be consider as part of island karst aquifer *i.e.* balance relation between salt and fresh water in according to the Ghyben-Hertzberg Low.

PROPERTIES OF THE VRANA LAKE SYSTEM

However, due to the specific properties of the Vrana Lake which has no directly measurable imflow nor run-off, the hydrological analyses of functioning of the lake system based on measuring of water balance parameters of the lake system (water level monitoring, precipitation regime, water pumping from the lake, evaporation from lake surface) have made a considerable contribution to the general level of knowledge on this natural phenomenon and its protection.

Due to its size compared to the comparatively small island enviroment, the lake containing 220 million cu.m. of potable water of exceptionally high quality affects significantly the dynamics of functioning of the entire aquifer. Except its position and size, the Vrana Lake is specific due to the fact that inflow into the lake, as well as the outflow, is accomplished by so far unidentified underground ways. No significant permanent springs have been noticed on the lake, except two temporary springs with the yield up to 0.005 cu.m./sec. Compared to the mater in the mainlend area beyond the direct influence of the sea, salinity of the lake water is somewat higher, ranging between 62 and 92 mgl-1. In the wider zone of the Vrana Lake - on the coast of the island of Cres several springs, vrulja and coastal sources have been noticed, as well as a zone of diffused fresh water outflow. Due to the constant outflow regime, some of these phenomena may be related to the Vrana Lake, or its aquifer, as possible privilleged directions of its discharging. Water level observations have been carried out on a daily basis since 1928, completed by rainfall observations on the catchment area and directly along the shore.

An expressed trend of water level decrease in the lake is noticed, in particular in the eighties. The trend of water level decrease in the entire period of analysis, 1929 to 1995, was 0,04 m per annum, while in the period from 1985 to 1990 it was even 0,48 m per annum. This trend caused the concern of the public and of the professionals for the destiny of the lake and possible disturbance of its balance, and possible breakthrough of sea water into its system. On the other hand, it may be seen that water pumping from the lake, which started in 1952, shows a clear growing trend, with the annual average for the 1964 to 1995 period of about 65 000 cu.m. per annum. During the recent decade, the average pumping rate from the lake is 0,072 cu.m./ sec., with the maximum during the summer season, up to 0,160 cu.m./sec. Fortunately, such drastic trend of water level decrease that could have endangered the balance of the fresh water aquifer of the Vrana Lake, was not continued - the rainfall conditions wereb improved, and due to the war situation in Croatia in the early nineties, the touristic traffic was reduced, resulting in reduced water pumpings from the lake, compared to pre-war maximum values.

RESULTS OF HYDROLOGICAL ANALYSIS

Multiple regression analysis of the annual changes of the lake status, in dependance on the magnitude of relevant hydrological parameters for the period from 1980 to 1995, when the most complete input data were available, gave a very acceptable regression dependance with the linear correlation coefficient k=0,96, wich may be seen. The resulting equation is: Hn=2,7369 On+1,12836 In– 4,0069 Cn+0,23208 Gn-0,31284 H(n–1)+0,3718 (1) Hn - annual water level change, On - annual rainfall, In-annual evaporation, Cn-annual pumping, Gn-calculated losses through sinking determinated on the basis of the sinking curve and H(n-1)-mean annual water levels in the previous year–all expressed in (m) with regard to the mean lake level. Such analyses, along with detailed analyses of sinking from the lake system and, provided the preconditions for elaboration of the complex mathematical model of functioning of the lake system of RANA which was used to compute the mean monthly inflows into the Vrana Lake. According to the computations, in the period from 1929 to 1995, the mean annual inflow into the Vrana Lake system was 0,588 cu.m./sec., and its value varied from 1,144 cu.m./sec. and 0,273 cu.m./sec.

The average value of inflow from the direct catchment area was calculated as 0,393 cu.m./sec., and figure of 0,195 cu.m./sec. refers to the rain falling directly on lake surface. The area of the orographic catchment area sourrounding the Vrana Lake oscillations and to the analysis of regional run-off, it comes out that the average catchment area satisfying the Vrana Lake recharging balance is only about 24 sq.km. This size of catchment area is not clearly determinable, changing due to the karst character of the terrain in dependance on hydrological conditions.

LAKE PROTECTION FROM OVERPUMPING

As already mentioned, the Vrana Lake is the only source of water supply of the islands of Cres and Loiknj, and at the present requirements the annual quantity of water pumped out of the lake reaches 2,3 million cu.m. As the calculated annual average inflows are 18.5 million, and the evaporation losses at the mean water level are of the order of 6,7 million cu.m., and the losses through sinking from the lake system are 11,7 cu.m.

it is obvious that pumping, although globally the smallest element of the water balance, disturbs the balance of the Vrana Lake system and influence lowering of the water level. This lowering of the mean lake level occurs to the case when the sinking losses are reduced in the magnitude of the increase of the pumped quantity. Namely, a functional relation has been determined between the quantity of sinking and the water level in the lake, and its gradient in domain of the observed water levels per each meter is approx. 0,028 cu.m./sec.

Until 1995, the total quantity of water pumped out of the lake reached 42 million m^3 which would, at the mean water level account for 7.3 m. The analysis showed that such water consumption, in spite of the expressed trend of water level lowering, caused, instead of the said 7.3 m, the actual lowering of the lake level by approximately 2 m. Therefore, it is obvious that this water loss was compensated by the change of hydrological conditions in the lake-first of all by reduction of the losses from the aquifer due to his lower level in relation to the sea, and partly due to the increased recharging from the underground part of the aquifer, caused by lowering of the lake level. The trend of recorded annual water levels in the Vrana Lake, and of values, obtained by mathematical simulation on the model of RANA, assuming there was no pumping, but also under the assumption that in the period of analysis pumping was even more intensive than the actual mean annual rate of 0,072 cu.m./sec., *i.e.* that they were double - 0,150 cu.m./sec., and 0,250 cu.m./sec., which is equal to the earlier forecasts of possible maximum allowed pumpings rates.

Althought the simulations of the Vrana Lakeb behaviour in cases of its intensified use have carried out on the basis of extrapolation of the determined regularities of behaviour in the real - higher water levels in the lake, and are therefore largely approximative, they still show that more intensive pumping of water from the lake may result in excessive consumption which, in turn, could cause disturbance of its dynamic balance with the sea. According to analysis minor planned increases of the pumping rate of the order of 20 to 30 per cent for the needs of the island population are not yet alarming. The formed fresh water lense of the island aquifer, by its higher pressure, on the principles of the Ghyben-Hertzberger Low, maintains balance with the sea and prevents its breakthrough into the lake. As the bottom of the Vrana Lake is even 61 m below the mean sea level, and with respect to the theoretical balance ratio (1m of fresh water lense above the sea level corresponds to 40 m of fresh water below the sea level) in that case the breakthrough of sea into the lake would occur at the lake level of 1,5 m a.s.l.

Such breakthrough into the lake would be a disaster for the lake system because the thermal currents in the lake, as well as winds, would cause salination of the entire lake, without any possibility of desalination later on. However, in reality, as the Vrana Lake is in the karst environment the limit of the contact of fresh water and the sea is not distinct, but involves a wider fresh and sea water mixing zone, which is considerably closer to the lake. Therefore, the above-mentioned theoretically possible lowering of the lake level should not be allowed, and increased pumping of water from the lake must be subject to constant control, and limited to the intensity that will not endanger the balance of the lake.

Vrana Lake is part of the complex system of island's karst aquifer, which functions in accordance with the principals of dynamical balance of the fresh water and the sea. In this paper are given the main characteristic of its hydrological system as also the analysis of problems of pumping from the lake. This paper shows that we can analysed aquifer behaviour by hydrological modeling when their natural hydraulic give limitation of excessively exploitation. It has been determined that the observed lowering of the water level in the Lake during the eighties was the result of the coincidence of extremely dry hydrological conditions and increased consumption of water from the Lake (pumping). Existing pumping of average 0,072 m^3/s does not disturb Vrana Lake to its exhaustion. Such development, in perspective, with unavoidable increase of water consumption standards, results in new requirements for water pumping out of the lake, thereby bringing new risks of disturbing the existing water quality.

In accordance with the conceptions we have found out it is obvious that the former estimations of permitted 0,250 m3/s were too optimistic. It has been established that rapid trend of falling of water level was caused by more intensive pumping from Vrana Lake and driver hydrological conditions.

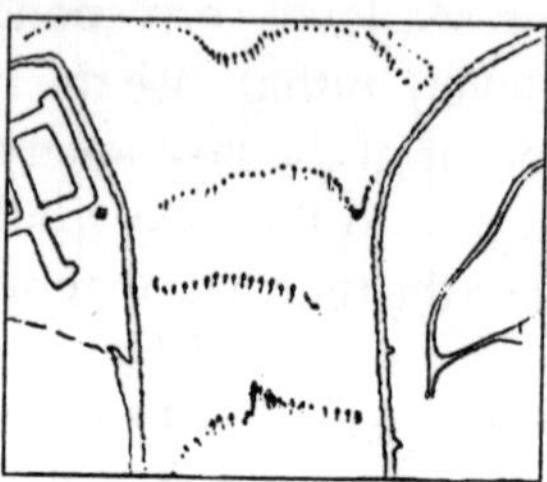

Fig. 8.1 Location Map

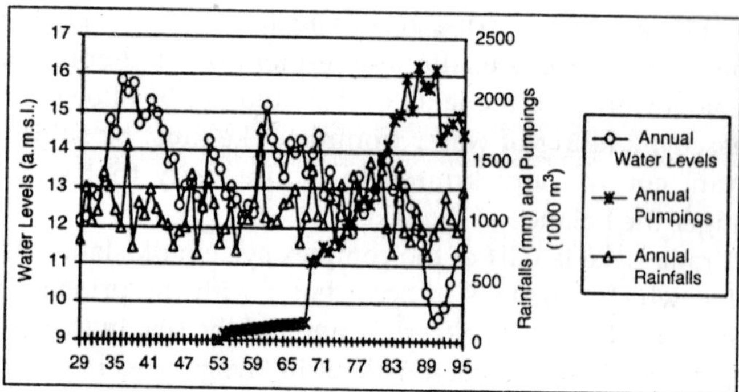

Fig. 8.2 Annual Values of Rainfall, Water Levels and Water Pumpings from the Vrana Lake

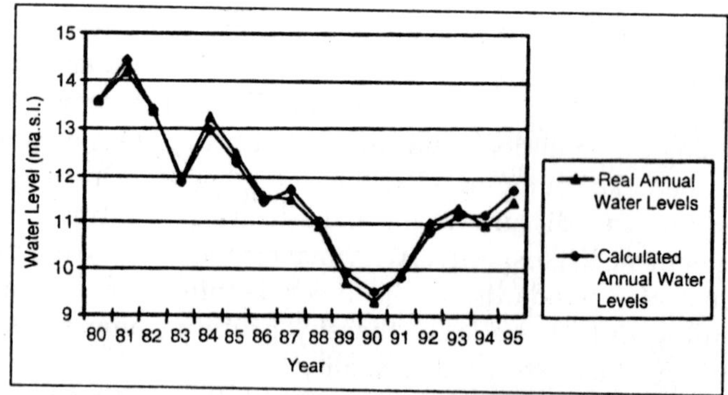

Fig. 8.3 Ratio of Recorded and Calculated Water Levels According to Multiple Regression Analysis

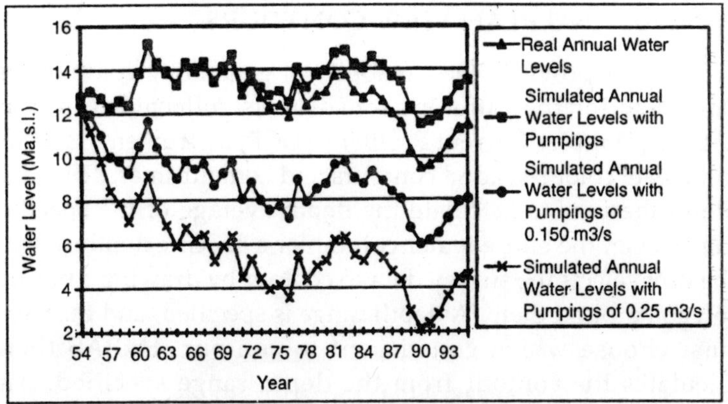

Fig.8.4 Annual Water Levels in Real Conditions, Withought Pumping, and With Pumping of 0,150 m^3s^{-1}, and 0,250 m^3s^{-1}

HYDRAULIC PROCESSES ANALYSIS SYSTEM

HyPAS builds on the inherent power of GIS while supplying easier tools for facilitation of hydraulic process analysis and reducing learning time typical with GIS implementation. HyPAS is an extension to ArcView, a commercially available software package marketed by Environmental Systems Research Institute (ESRI), and requires an additional extension from ESRI, 3D Analyst. HyPAS was designed for the non-GIS expert with ease of use as a priority. The HyPAS was designed to perform all major functions after collection of data to report writing.

HyPAS's velocity analysis tools cover three basic applications: contouring an area in plan view from a user-defined constituent and depth range, generating cross sections from a transect, and plotting vector magnitude and direction in plan view from a user-specified depth range. HyPAS's sediment sample analysis tools allow the user to generate frequency weight plots, calculate composite sample plots, and perform varied analysis routines. HyPAS's project management tools allow the user to import photographs for project enhancement and import time series data to manage and plot.

GENERATING PLAN VIEW CONTOURS

HyPAS provides the capability of generating contours in plan view for the different constituents collected from the Acoustic Doppler Current Profiler (ADCP) instrument. HyPAS will create colour-shaded contours and contour lines. This tool allows the user to calculate the depth average of the specific values from the entire water column or a specific depth range. The user selects the survey data to contour by drawing a box or polygon around them. A depth range is specified, and the user must choose which constituent to contour. HyPAS then calculates the contour from the depth range specified. An example plan view contour of total velocity magnitude was created from ADCP data in the Columbia River, Oregon. The yellow dots represent actual velocity profiles collected by the ADCP as the boat moves along the line. The legend for the colour-shaded contours represents velocity magnitude in ft/sec. The units for velocity magnitude depend on how the data were processed. Often, velocity magnitude is represented in cm/sec.

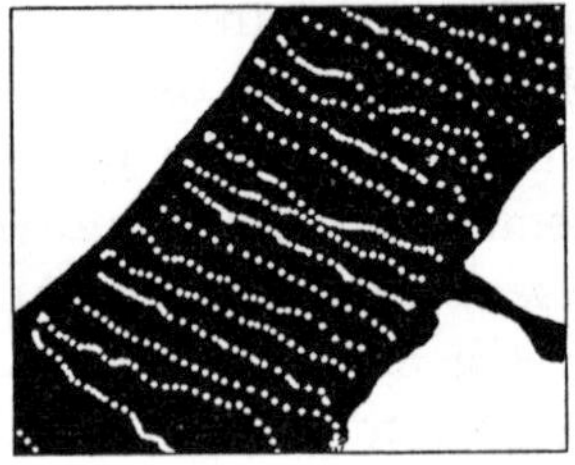

Fig. 8.5 Plan View Contour of the Total Velocity Magnitude from an ADCP Survey in the Columbia River

GENERATING CROSS SECTIONS

HyPAS provides the capability of generating a cross section from a specific transect or points along multiple transects from velocity survey data. This tool allows the user to calculate a cross section from any constituent in the data. This includes the north, east, and vertical components of velocity as well as acoustic backscatter. The user selects the survey data points by drawing a polygon around the individual points and selecting a starting point. The constituent to contour is selected. HyPAS then interpolates and plots the cross section. A cross section was created from ADCP data collected from an area called Victoria Bend in the Mississippi River.

IMPORTING AND DISPLAYING IMAGES

HyPAS provides the capability of importing and displaying digital photographs and images to enhance project management. This tool is not designed for image processing, but rather as an end product aid for representing analysis results and methods. This tool allows the user to import digital photographs and images and attach them to specific spatial locations throughout the study area.

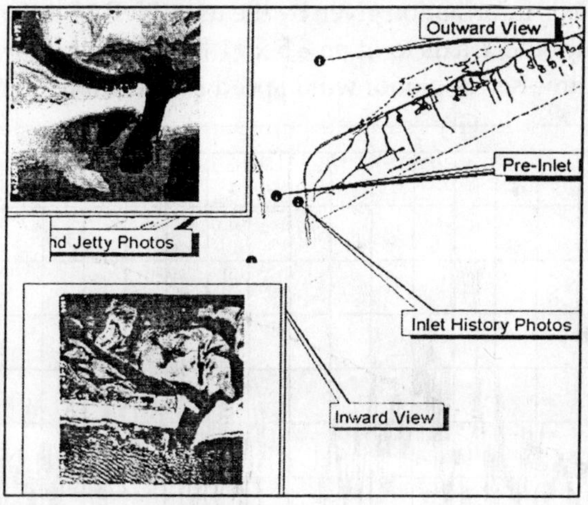

Fig. 8.7 Shinnecock Inlet GIS Project Utilizing HyPAS Tools for Displaying Photographs

The images are stored using a separate image theme as shown in Figure 8.7. On the image theme, a small solid filled circle with an "i" denotes the locations of the images. Presently, a maximum of ten digital images can be stored at each spatial location. This tool is useful for recording conditions at a site or progress of a construction site. Images allow the user the opportunity to store and display instruments, flow conditions, historic photos, and other visual information quickly and easily for reference. Several photographs were added to a GIS project for Shinnecock Inlet, New York. Two photographs are displayed showing the inlet at previous times in its history.

TIME SERIES ANALYSIS

HyPAS provides capabilities for importing, storing, analyzing, and exporting time-series data. This is a project management tool that allows the user to have time-series data flags throughout his project. The user create a location for time–series data and import the data for that location. Later the user can choose that location and HyPAS will display all the data types, which have been imported. HyPAS will plot the data on an X, Y plane using the plot and axis information given by the user. HyPAS automatically exaggerates the Y axis to fit an 8.5 x 11 landscape plot. Figure 8.8 shows a time-series plot of wind speed information.

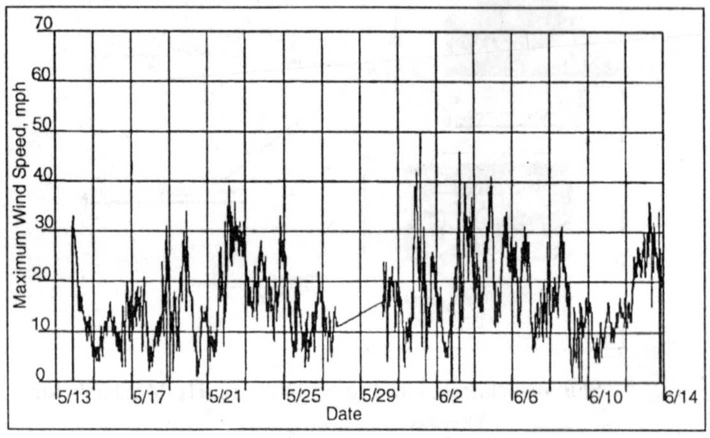

Fig. 8.8 Maximum Wind Speed Time-Series Data Plot

PLOTTING GRAIN SIZE DISTRIBUTION

HyPAS provides the capability of plotting grain-size distribution for sediment sample data. The user can plot cumulative frequency weight or frequency weight per cent histograms. HyPAS can plot all samples selected or calculate a composite for the selected samples. The user has the option of sorting the tabular data before making a selection of samples for display in the plotting window.

Once sediment samples are imported, the user must select the samples desired for a grain-size distribution plot before selecting the Sediment Sample Tool. The option to plot the cumulative frequency weight percents, the frequency weight per cent histogram, or a composite sample plot is provided. HyPAS then generates the distribution plot. Figure 8.10 shows frequency weight per cent histograms with a composite histogram from a subset of sediment samples collected from the Shinnecock Inlet.

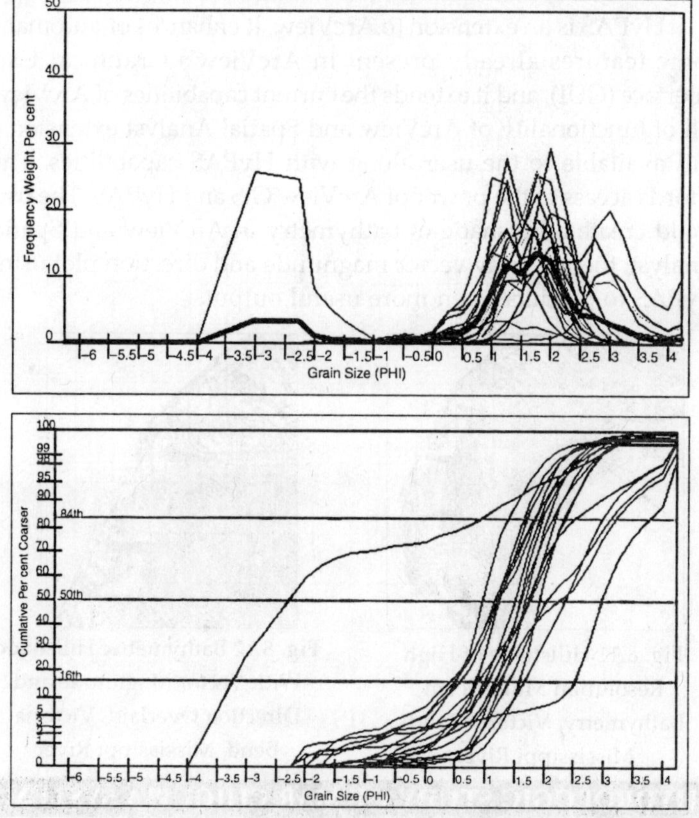

Fig. 8.10 Example of a Series of Sediment Samples and Composite Sample.

The First Plotdisplays the Frequency Weight Per cent Histograms, and the Second Plot Displays the Cumulative Weight Per cent.

A cumulative frequency weight per cent plot of the same sample is also shown. The thicker line delineates the composite sample. Various statistical parameters describing this composite sample are stored to a table. These statistics include the median, other percentiles, the Fill Factor (RA), and the Renourishment Factors (RJ). Different composites can then be easily compared.

EXTENSION INFORMATION

HyPAS is an extension to ArcView. It enhances or automates some features already present in ArcView's Graphical User Interface (GUI), and it extends the current capabilities of ArcView. All of functionality of ArcView and Spatial Analyst extension is still available to the user along with HyPAS capabilities. This affords access to the power of ArcView GIS and HyPAS. The user could create a hillshade of bathymetry in ArcView and Spatial Analyst; then overlay vector magnitude and direction plot using HyPAS to produce even more useful output.

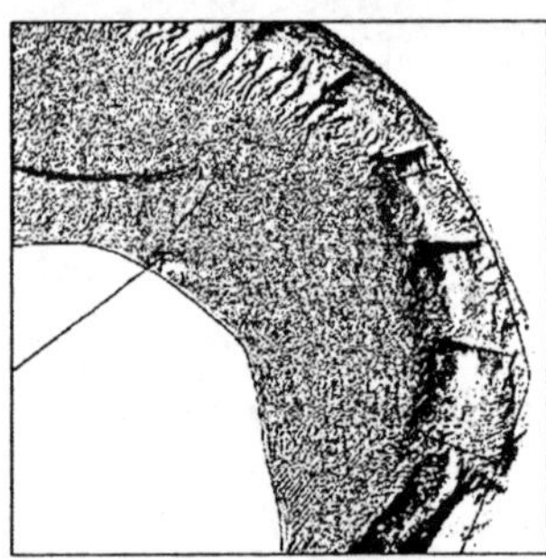

Fig. 8.11 Hillshade of High Resolution Multi-Beam Bathymetry, Victoria Bend, Mississippi River

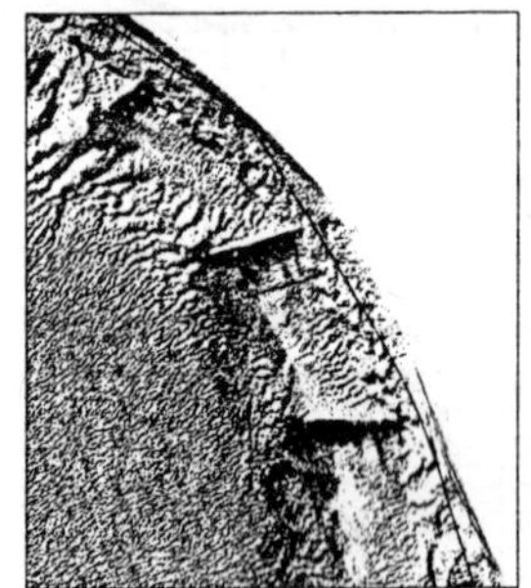

Fig. 8.12 Bathymetric Hillshade With Vector Magnitude and Direction Overlaid, Victoria Bend, Mississippi River

HYDROLOGIC STUDY OF MAR CHIQUITA SYSTEM

Mar Chiquita and its wetlands are located at the center of Argentina. It is a large and complex eco-hydrologic system mainly formed by a main salt lake, lagoons, wetlands of Dulce River in the Northen area, and by the semi-ephemeral Suquía and Xanaes rivers. It is the biggest salt lake in South America, with an area near 6.000 km^2, characterized by its little depth (approximately 10 m.) and large wetlands of 10,000 km^2 with a high biodiversity. Numerous hydro-meteorological information have been compiled, which consist of series on precipitation, evaporation, temperature and wind in Córdoba and Santiago del Estero provinces. Series of levels and flows in the main tributaries have been obtained, which were also filled for the simulation period. The water level series of Mar Chiquita lake were obtained at Miramar City (south coast).

The geometric functions of the lake (curves height-area-volume) were computed using previous bathymetric data up to 66 m over sea level and a set of 8 satellite images over such level. In this way, a set of consistent information has been completed, allowing a six-month step hydrologic balance within the last 30 years covering three types of climatic periods (dry, medium and wet). The results of the numeric simulation allow a better knowledge of this complex system supporting a solid basis for the water resources management in this region.

Being of special utility to analyse the effects that can generate the increase of irrigation intakes from Dulce River upstream wetlands and lake, in order to prevent reduction effects such as the Aral sea case.

HYDROLOGIC SIMULATION OF MAR CHIQUITA SYSTEM

The conceptual model of Mar Chiquita System updates previous studies including the following hydrologic components: Mar Chiquita lake body, Dulce river wetlands area, three tributaries: Dulce, Suquía and Xanaes rivers from field measurements, Precipitation, Evaporation and flow extractions in the tributaries. The numerical model of the hydrologic balance solve the conservation of mass equation applied the system,

$$\Delta V = \Sigma\, Q_i\, {}_{\Delta}T + A_L\,(P\text{-}E)\,\Delta T - Q_{ext}\,\Delta T$$

Being Δ V volume variation of the lake, Q_i flow contribution of the tributaries, P media areal precipitation over the lake, E evaporation, A_L lake area, Q_{ext} flow extraction to the system and ΔT time step. The model was initially calibrated by adjusting the series of the tributary contributions, measured upstream the lake to the balance results assuming as data the series of precipitation, evaporation and lake levels. The extraction flows to the system were assumed to be void during the process of model calibration, and variables between 4 y 60 m^3/s during the simulations. The time step was, according to the simulated case, 6 or 12 months. The results shown herein are with a 6 months interval. The calibration was performed for the 1967-1997 period, due to the availability of 30 years with all basic data for the calibration of model (mainly mean water levels).

Evaporation

As evaporation estimators, Lungeon, Meyer and Rohwer were compared with measurements, selecting the first one (ec. 2), to which a correction for the occurrence of variable salinity was added:

$$Em = 0.398.\ D.\ (es-ed).\ (\ (273-T)/\ 273\ ).\ 760/\ (Pa-es)$$

Being: Em monthly evaporation, D number of days per month, es and ed: monthly steam saturation pressure average and monthly steam pressure average respectively (mmHg), T monthly average value of the maximum daily temperatures (°C), Pa mean atmospheric pressure (mmHg). The saturation steam pressure was calculated as es = ew-0.00066.Pa.(Ta-Tw).(1+0.00115.Tw), with ew, steam pressure corresponding to the average temperature with humid bulb thermometer,Ta air temperature and Tw average temperature with humid bulb thermometer.

The monthly average steam pressure was estimated as ed= HR.es/100, being HR the relative humidity (%). The results of vertical balance of mass in the lake and wetlands have been published by Pagot el al. The volume variations have been obtained from the measured series of water levels and the corresponding geometric functions Area (level) and Volume (level).

Levels of Mar Chiquita Lake

The water levels were measured in the south coast of the lake in Miramar City. These series were previously corrected by datum changes. In this research monthly levels were adopted to avoid the influence of the wind effects ("wind set up" that can affect in more than 0.5 m). As it was mentioned before, the availability of measured levels from 1967 to 1997 has allowed the pre-calibration of the model in this period, when closing the equation for the volume balance because all its terms were known.

The Geometric Functions

They were obtained from the bathymetric campaigns carried out by CIRSA completed with 6 satellite images in the periods corresponding to different water levels of the lake. The satellite images used correspond to the following years: 1972, 1976, 1981, 1986, 1993 and 1997, covering the periods of low, medium and high water levels. Images corresponding to the years: 1976 and 1981 are shown in figure 8.13.

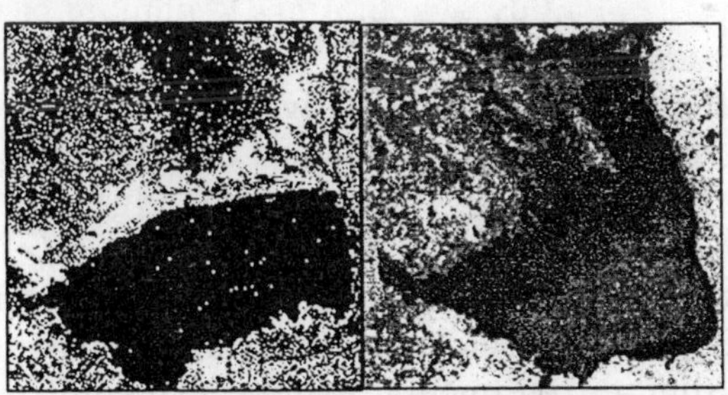

Fig. 8.13 Satellite Images of Mar Chiquita Lake for Medium -1976- and High Levels -1981.

Precipitation

The contributions corresponding to the precipitation have

been obtained from the pluviometric series, generated in eight sub-series, which correspond to each side of the lake (N, S, E, W, SE, NE, NW and SW). These ones were formed in turn from series measured by DIPAS in 23 pluviometric stations. The calculus of average areal precipitation considers the relative influence of each sub-series due to the lake area through an expression of the form,

$$P= \alpha_N (h)P_N + \alpha_{NW} (h)P_{NW} + \alpha_{NW} (h)P_{NW}\alpha_S (h) P_S + {}_E(h)P_E+\alpha_O (h)P_W+\alpha_{SE} (h)P_{SE}+\alpha_{SW} (h)P_{SW}$$

With $\Sigma\alpha_i = 1$, where the coefficients obtained by means of Thiessen polygons.

Salinity Variations

As mentioned before, the evaporation was obtained from direct measures taken in Miramar tank (with a correction factor = 0.7), and the calculus by Lungeon expression using meteorological data measured in the periods without direct evaporation measures. In the intervals without meteorological measures equivalent series were used, generated from the monthly average values measured in Miramar. The correction of the evaporation due to salinity variations was determined by,

$$EVP\ c=EVP\ (2-\rho)$$

where is the lake water density, which is calculated by the following empiric expression obtained from the measured

data: $\rho = 0.0007\ S + 1$

where S is the total content of salt dissolved in the water of the lake expressed in gr/l. This content of total dissolved salt S was obtained from a experimental relation that combines direct measures of S and the volumes V corresponding to the sampling dates. These functions were included in the model to automatically carry out the direct and inverse transformations.

RIVER FLOWS

The monthly flows near Mar Chiquita lake for Xanaes and

Suquía rivers have been obtained from series of water levels and some discharge measurements. The rating curves were determined near the lake, extrapolating experimental values with the Manning formula, assuming steady and uniform flows. Dulce river discharges into the lake are one of the main results of the model, due to they can not be measured near the Mar Chiquita because of the wetlands existence and the very difficult access conditions. Then the closer data have been measured in Los Quiroga dam, near 300 km upstream the lake, where the main present and future extractions for irrigation are located.

Numerical Simulations

The numerical model is a simple dynamic one, developed in Fortran to solve equation 1, that applies a time discretization in finite differences and solves the non linear terms by means of an interactive standard procedure until reaching the convergence (*e.g.* the evaporation due to salinity, which is dependent of the volume that is unknown in each temporal step). The empiric functions of direct geometry A(H), V(H) and the inverses H(A) and H(V), and those of salinity S(V) and V(S) were conveniently adjusted as variable degree polynomials by sections, being evaluated in high precision in order to reduce the numeric error. The simulations were carried out with the model (previously pre-calibrated for the period between 1967 and 1997), assuming constant flow extraction in each one.

This can be seen in the reduction of lake volumes and the increase of salinity levels. However, for these small extractions, the lake-wetlands system demonstrates the capacity to quickly increase its dimensions if the hydrological conditions change, when increasing the precipitations and surface contributions, like it happened from 1976 when the lake overcame the average levels (66 m o.s.l.). For extractions of major flows -*e.g.* 10 m^3/s - the critical interval of the lake increases in almost one year. This critical period occurs when salinity level is greater than 50 gr/l - considered the first threshold of salinity for fish reproduction-.

It is also observed that with these extraction flows, at the beginning of the 90s and during more than 3 years the levels of salinity overcome the 50 gr/l. For extractions of 20 m^3/s the recovery of the lake would take 3 more years, up to 1981. Also, from 1988 salinity values would severely increase, overcoming a second threshold of salinity of 100 gr/l (a limit for fish survival). At the end of the period the level of salinity would stay above of the critical value without never ending up recovering to smaller values or similar to 50 gr/l. For major extractions the effects would be extremely severe. The lake would dry off in more than half of the period simulated of 30 years, and salinity, would always stay above the first threshold and in approximately 80% of the time above the second threshold.

The modelling of Dulce river discharge showed a flow reduction near 40% from Los Quiroga dam to the lake, 300 Km downstream, due to the effects of the wetlands which act as a natural reservoir increasing evapo-transpiration.

MODELLING THE WATER BALANCE OF NATURAL LAKE

INTRODUCTION

In the Southern part of Rift Valley Region of Ethiopia there are two natural Lakes, namely Abaya and Chamo Lakes, and little information is available about these Lakes. The watershed of the Lakes is about 18,000km^2 including water area and currently inhabited by more than 3 million people with growth rate of about 3% per annum. These Lakes are important shallow Lakes and among others importance, to inhabit a number of aquatic animals and to modify the climate of the area.

Yet there are evidences that they are heavily influenced by human activities and endangered Lakes unless proper precautionary measures are undertaken in the Lake's watersheds. In order to investigate the available water resources quantity in the lakes as well as impacts and influences of natural and man-induced factors a water balance model of the two Lakes have been developed. The knowledge of water balance of lakes and reservoirs is an essential component of water management. Water management decision as far as possible should be based on a thorough quantitative understanding of the hydrologic cycle of the lakes/reservoirs in the basin. In developing country like Ethiopia, the quantitative understanding of individual components of the hydrological system themselves is difficult. This paper focuses on developing a water balance model, and describe shortly methods employed to obtain the physical characteristics of the watersheds and the lakes as well as the components of hydro-meteorological elements of the water balance components under limited data situation.

The developed information is used for the simulation of lake water level. Furthermore, impacts of various influences such as sediment and water use are described and expected life of the Lakes under sediment load are discussed and measures to reverse the deteriorating situation are briefly discussed.

MODEL EQUATIONS AND SOULTION PROCEDURE

Basic Equations

The input and output components of the water balance of a lake or reservoir depend not only on the physical dimension of the water body, but also on the climatic, hydrological and geological factors affecting the water body and its surrounding areas. The water balance equation can be written, from continuity equation at any time, which is governed by the conditions that the water volume remains constant. The continuity equation intern governed by conservation of matter, which described by equilibrium between added water volume or depth, lost water volume or depth and change in volume or depth as:

$$V_{in} - V_{ou} + P - E - \Delta S = 0$$

Where Vin is surface and subsurface inflow; Vou is surface and subsurface outflow; P is precipitation volume; E is evaporation volume; S is change in storage. Alternatively, parameters can also be similarly defined in terms of depth of water. In ideal situation variables of the water balance equation are computed separately, and providing closed result. In practice however, the computation leads to a discrepancy or residual error. Considering the error term, d, the above equation can be re-written as:

$$V_{in} - V_{ou} + P - E - \Delta S \pm \delta = 0$$

In the above equations parameters can be distributed as:

$$V_{in} = V_{si} + V_{ssi}$$

$$V_{si} = \sum_{i-1}^{ngw} V_{g} + \sum_{i-1}^{nuw} V_{ug}$$

$$V_{ou} = V_{so} + V_{sso}$$

Where respectively, V_{si} and V_{so} are sums of surface inflow and outflow; V_{ssi} and V_{sso} are sub-surface inflow and outflow; V_{g} are V_{ug} are gauged and ungauged inflows; ngw and nuw are number of gauged and ungauged watersheds. The error term, δ, is treated component wise.

Solution Procedure

The water balance model equations written in the above form, using various water balance components can be used to compute and simulate the water volume, area, depth or alternatively unmeasured water balance components. The intention of the water balance in here is to simulate the water level and compute volume, area and their temporal variability on monthly or yearly time spans. In order to simulate the water level, volume based or depth-based equations, which have equal applicability, can be employed.

The depth based simulation procedure has been employed for results in this paper and described by the following steps:

- Compute initial parameters such as area and volume from the initial depth as boundary conditions,

- Assume, mean area in time period i, $A_{mi} = A_{1,i}$, where A1,i initial lake area,
- Compute change in depth, ΔZi,, from

$$\Delta Z_i = \frac{(PF \cdot p - EF \cdot e) \cdot A_{inj} + V_{inj} - V_{ouj}}{A_{mj}}$$

$A_{l,i}$ is area of the lake for evaporation and rainfall computation, with p and e are rainfall and evaporation depths respectively. PF and EF are rainfall and evaporation correction factors, which are used to adjust rainfall and evaporation on the lakes.
Depth at the end of the simulation interval is:

$$Z_{2,i} = Z_{ij} + \Delta Z_i$$

- Compute $A_{2,i,}$ the area at the end of time interval i from area elevation curve,
Compute $A_{m,i,}$ from:

$$A_{m,i} = \frac{A_{l,i} + A_{2,j}}{2}$$

- Repeat steps 3 & 4 until reasonable agreement in Z obtained,
- Compute, volume $V_{2,i}$, at the end of time interval, i, using capacity curve,
- For the next time interval $Z_{1,i+1}$, $A_{1,i+1}$, $V_{1,i+1}$ are described by $Z_{2,i}$, $A_{2,i}$, $V_{2,i}$.

INPUT DATA PROCESS, STRUCTURE AND DATA

Structure of Data Elements

The programme for water balance named, LAKEBAL, has been developed and the modelling process constitutes three stages:

- Pre-processing of input data and selection of model parameters
- Simulation and computation of water balance components by executing the simulation programme
- Post processing of output from simulation result

Data of Water Balance Components

The components of the water balance, such as run-off, rainfall and evaporation, lake morphometry and etc., were not existing and these components have been developed in the wider scope of a research associated. The data can be stated as watershed data, lakes morphometric data, meteorological data, run-off data and peculiarities in hydrology of the watersheds.

The composed data for hydro-meteorology totals 27 years of Abaya and 22 years of Chamo Lake on monthly basis. The watersheds of the two lakes drainage basin have been modelled combining ArcView GIS and Watershed Modelling System (WMS) hydrological model. While the former has been employed for Digital Elevation Modelling (DEM) the latter has been used for Digital Terrain Modelling (DTM) in derivation of extensive hydrological parameters, such as slope, basin area, perimeter, etc.

After refining the sub watersheds capturing the locations of interest such as gauge stations and lake outlets, the entire basin was subdivided in to 52 sub-watersheds. The watershed data under the GIS environment masks the Lake morphometry due to data unavailability. In order to close this gap, a bathymetry surveys of the two Lakes have been undertaken combining Global Positioning System (GPS) and Echo-sounder.

Most of the area in the two Lakes basin is ungauged (only 43.9% of land area is gauged). While the data of gauged areas have been used, a rainfall-run-off model based on monthly water balance concept was developed, to estimate ungauged portion of the watersheds as well as the missing elements of data in the simulation period.

Components of the subsurface flows are accounted via surface modelling and direct accounts are disregarded. Certain peculiar characteristics regarding the hydrology of the watersheds such as intensive evaporation due to wet land and inundation areas, hot springs, water uses in irrigation are accounted as these elements affect the magnitude of inflow components. Furthermore, surface outflows are accounted in the simulation model from the gauged information or by trial and error where such information is unavailable. The existing Lake level records constitute three gauge stations on Abaya and one station on Chamo Lake. While the Abaya Lake data could be filtered, corrected and missing values are filled by comparing the stations, in the case of Chamo Lake data certain years of monthly uninterrupted records at the beginning and end of simulation periods could be used, because the intermédiate values are associated with shift of stations and change of recording mechanisms.

CALIBRATION OF PARAMETERS AND SIMULATION RESULTS

Calibration

The EF and PF parameters are used as calibration parameters. Optimum values of these parameters can be selected by minimising the error terms between measured and simulated Lake water level using ordinary least square procedures or by visual inspection of plotted results combined with regression equations. The optimum values of the calibration parameters can be set for various scenarios of investigating impacts on the lakes.

Simulation Result Using Lakebal

Due to uncertainty on the data of the water balance components on one hand and the need to obtain acceptable accuracy between the simulated and measured parameters on the other hand, the simulation and comparison of results were carried out for various scenarios.

Initial Simulation Without Considering Outflow

The first simulations have been made for both Lakes without considering outflows. After a number of trial simulation, it is shown in Figures 8.4 and 8.5 with respective description the simulation run of Abaya and Chamo for no outflow conditions, under the best selected parameters which suites simulation of level before commencement of outflows.

Simulation Considering Outflow

The simulation considering outflows are converged with EF parameters of 1.83 and 1.75 and PF parameters of 0.85 and 0.90 respectively for Abaya and Chamo Lakes. The higher evaporation factor for Abaya Lake is caused by higher heat absorption capacity of the Lake, underestimation of the method of evapotranspiration as well as uncertainty of the data derived from lake periphery. The lower rainfall factor is due to possible pluviometric depression effect of the rainfall falling on the Lakes.

The coefficient of determination of the regression of measured and gauged values for Abaya and Chamo Lakes under these conditions are 0.87 and 0.99 respectively. Analysis of error shows, the simulation result is highly sensitive to local error, and as such if an error due to data occurs at a particular point, the error propagates in the subsequent simulation periods and shows apparent errors.

Simulation Considering Estimated Sediment

The effect of deposited sediment on the lake is clear and is detrimental to the water carrying capacity of the Lakes. The sediment delivered in to the Lakes has no outlets and fully deposited. Sediment load entering the Lakes particularly that of Abaya is high in magnitude. The causes of the large quantity of sediment are associated to intensified erosion in the catchment area due to deforestation, overgrazing and poor farming practice.

In terms of Lake level simulation, the effect of inclusion of sediment deposition is to loose the bottom capacity of the Lakes and thereby an increased water level. In order to quantify the effects of sediment, unfortunately data of neither monitoring of sediment transported and delivered in to the Lakes nor adequate sediment measurement at some points in the rivers systems is available. In order to assess impact of sediment on the Lakes, 2 pairs of three various conditions of sediment inputs were evaluated. These are yield or concentration based under tolerable sediment, moderate and high sediment rates. These rates are based on watershed yield based to be 0.1, 2 and 25 t/ ha per year respectively and the concentration based to be 0.02, 4 and 25 kg/m³ respectively (scenarios II-1, II-2, II-3). Based on the various scenarios of sediment input the water level simulation and the life expectancy of the Lakes were evaluated.

Under this case, the existence to complete disappearance of Abaya Lake is estimated at 207 years and that of Chamo is estimated at 467 years. The latter would be dramatically reduced once Abaya is filled with sediment and the existence of Chamo would be then few decades after Abaya's disappearance. Investigation of research catchments located outside the watershed of the Lakes, see Dawit and other rivers in Ethiopia which have similar watersheds and climatic characteristics as well as evidences of the sediment load (short period sampled data) in the tributary rivers of the Lake show the sediment input in the two Lakes is rated medium to high rate. As a result the two shallow Lakes are threatened by the large deposition of sediment.

Index

A

B

C

D

E

F

G